ALLE ZEIT WACH
1842

Heinz Engelhardt

High Performance Liquid Chromatography

Chemical Laboratory Practice

Translated from the German
by George Gutnikov

With 73 Figures and 13 Tables

Springer-Verlag
Berlin Heidelberg New York 1979

Heinz Engelhardt
Angewandte Physikalische Chemie, Universität des Saarlandes,
D-6600 Saarbrücken, Germany

George Gutnikov
California State Polytechnic University, Pomona, CA 91768, USA

Enlarged and revised translation of the second edition of *Hochdruck-Flüssigkeits-Chromatographie* by H. Engelhardt in the series *Anleitungen für die chemische Laboratoriumspraxis,* Vol. 14 (Springer-Verlag 1977)

ISBN 3-540-09005-3 Springer-Verlag Berlin Heidelberg New York
ISBN 0-387-09005-3 Springer-Verlag New York Heidelberg Berlin

Library of Congress Cataloging in Publication Data. Engelhardt, Heinz, 1936 –. High performance liquid chromatography (Chemical laboratory practice). "Enlarged and revised translation of the 2nd edition [1977] of Hochdruck-Flüssigkeits-Chromatographie." Bibliography: p. 1. Liquid chromatography. I. Gutnikov, George, 1938 –. II. Title. III. Series. QD117.C5E5313. 544'.924. 78-22002

Printed in Germany

Printing: J. Beltz, Hemsbach. Binding: Konrad Triltsch, Würzburg
2152/3140-543210

Contents

Foreword

Modern liquid column chromatography (LC) has developed rapidly since 1969 to become a standard method of separation. If the statisticians are to be believed, the recent growth of LC has been the most spectacular development in analytical chemistry and has not yet abated because its vast potential for application remains to be fully exploited. Significant factors contributing to this continued rise are the simplicity and low cost of the required basic equipment and the relative ease of acquiring and interpreting the data.

Unfortunately, in LC, as so often in the field of analytical chemistry, the available commercial instruments are frequently far more complicated - and consequently far more expensive - than is necessary for routine application. Therein also lies the risk of propagating a "black box" philosophy that would be particularly detrimental to chromatography. Moreover, it appears to have been forgotten, as was done previously with gas chromatography, that inadequate separation by a column can be remedied only with great difficulty, if at all, by electronic means. Also, whether the capillary columns recently advocated with great enthusiasm for LC will fulfill the expectations of their proponents is highly questionable unless someone comes up with some new and revolutionary ideas.

Of course, the most complex separations will still demand the latest state-of-the-art equipment. But surprisingly little evolution in column technology has taken place in recent years. From the literature it appears quite evident that the overwhelming majority of today's actual separations could be achieved equally well under less stringent conditions - on shorter columns with smaller pressure drops (< 50 bar) - and, hence, with less sophisticated instrumentation. The ready availability of such equipment should contribute significantly to wider use of high performance liquid chromatography (HPLC) as a routine analytical tool.

It is to be hoped that this book will enable the reader to appreciate these more pragmatic aspects of chromatography and thus help him to select the proper approach and suitable apparatus for solving his particular separation problems.

Saarbrücken, August 1978 I. Halász

Preface to the English Edition

The primary objective of this book is to present modern (high performance) liquid chromatography in a simple, non-mathematical manner, both for the beginner and the seasoned practitioner. Accordingly, the theory is kept brief, and only quantities that are easy to measure or can be taken directly from a chromatogram are included. Moreover, an attempt is made to express the theoretical concepts, which frequently emerge as equations, in readily comprehensible language and to illustrate them with appropriate examples. Considerable detail is devoted to factors that govern or affect chromatographic separations, and to the available means of manipulating them to achieve optimum results. Also stressed are the factors that may lead to errors, misinterpretations or poor reproducibility; this should not suggest, however, that this technique is less reproducible than other chromatographic methods.

This English version is based on the second German edition, but an effort has been made to update the literature thoroughly without excessively enlarging the book.

This book contains the results and ideas of long and fruitful discussions within the research group of the Institute of Applied Physical Chemistry. I am grateful to my colleague, Prof. I. Halâsz, as well as the co-workers of our research groups for their cooperation, suggestions, and patience. The final, updated version of this translation was completed in close collaboration with Prof. G. Gutnikov during his sabbatical leave spent at Saarbrücken.

Saarbrücken, August 1978 H. Engelhardt

Acknowledgements

The permission granted to reprint the following figures is gratefully acknowledged:

to Analytical Chemistry (Copyright American Chemical Society, Washington, D.C.),

to reprint Figures: VI.19; VI.22; VI.25a,b; VII.1; VII.2; VII.9; VII.10;

to Marcel Dekker Inc., New York, N.Y.,

to reprint Figure: VI.7;

to Journal of Chromatography (Copyright Elsevier Scientific Publishing Company, Amsterdam, The Netherlands),

to reprint Figures: VI.5; VI.6; VI.16; VI.18; VI.26; VI.27; VII.4; VII.5; VIII.2; IX.4;

to Wiley-Interscience, New York, N.Y.,

to reprint Figure: VII.3.

Our appreciation is also expressed to the companies involved in HPLC for their permission to include figures from their descriptive literature in this text.

Chapter I

Chromatographic Processes

Chromatography with a liquid mobile phase can be traced back to the work of the Russian botanist, M. Twsett [1].

Initially, chromatography was carried out with columns that were more than one cm in diameter [2-5]. This method was used for qualitative and quantitative analyses as well as preparative separations. The flow of the mobile phase through the column packing was effected by gravity, and was occasionally accelerated by hydrostatic pressure. In spite of this, the flow rate was less than 60 ml/h per cm^2 of column cross-section (corresponding to a linear flow rate of less than 0.02 cm/sec). The average particle diameter had to be around 100 μm or larger to achieve these flow rates with the available hydrostatic pressure. Because of the large particle size the efficiency of such columns was not particularly good, another reason probably being the constant overloading of the column with sample. The planar methods introduced later, namely paper [6,7] and thin layer chromatography [8,9], extensively supplanted the classical column technique. In most cases the separations were better and faster than in columns, and the identification of the separated samples using spray reagents was considerably simpler.

In the last few years column chromatography with a mobile liquid phase has undergone a renaissance, in part as a result of the development of sensitive detectors for the sample components in the column effluent, and also from the transfer of the accumulated knowledge of gas chromatography to liquid chromatography [10].

In high pressure (performance) liquid chromatography, HPLC, (sometimes called high-speed liquid chromatography) narrow columns with internal diameters of 2-8 mm are used. These columns are packed with particles having an average diameter of less than 50 μm. The velocity of the mobile phase is increased by means of a high inlet pressure (10-400 atm), and in general, a linear flow rate between 0.1 and 5 cm/sec or even higher is utilized.

Classification of Chromatographic Processes

Every chromatographic separation is based on differences in the rates of migration of the sample components through the column. The different sample components spend different times in the stationary phase, whereas the time spent in the mobile phase is identical for all components. The mobile phase, which alone effects transport through the column, can be either a gas or a liquid. Classification is based on the type of mobile and stationary phase used. If a solid with a large active surface is employed as the stationary phase and a gas as the mobile phase, we speak of gas-adsorption chromatography (or gas-solid chromatography, GSC). With a liquid mobile phase we have liquid-adsorption chromatography (or liquid-solid chromatography, LSC). If the stationary phase is a liquid coated on an inert support that has a large pore volume, it is classified as gas-partition chromatography (gas-liquid-chromatography, GLC) or liquid-partition chromatography (liquid-liquid-chromatography, LLC).

If a solid is used as the stationary phase, one generally speaks of *adsorption chromatography*, if a liquid is coated on an inactive solid support, it is called *partition chromatography*. No clear differentiation can be made between the two. Especially in adsorption chromatography where precoated adsorbents are frequently used, there is a continuous transition from pure adsorption to more or less distinct partition [11]. Similarly, for partition processes the influence of the support on retention cannot be neglected, particularly if its surface area is large.

This classification is unambiguous only in the case of ion-exchange and exclusion chromatography (also called gel filtration or gel permeation). Ion-exchangers [12,13] are insoluble porous materials (nowadays mostly organic polymers) having cationic or anionic sites at the surface that can exchange anions or cations from the mobile phase. In exclusion chromatography [14,15] porous solids with a defined narrow distribution of pore sizes are used. Molecules with effective diameters greater than those of the pores cannot diffuse into the interior of the solid and are thus more rapidly transported through the column than the smaller molecules that penetrate the pores. This is true because there is no transport within the pores in the axial direction of the column.

Gas-chromatographic methods are appropriate for the separation of volatile substances. Many substances that are nonvolatile at ordinary pressures can be converted simply and quantitatively into volatile de-

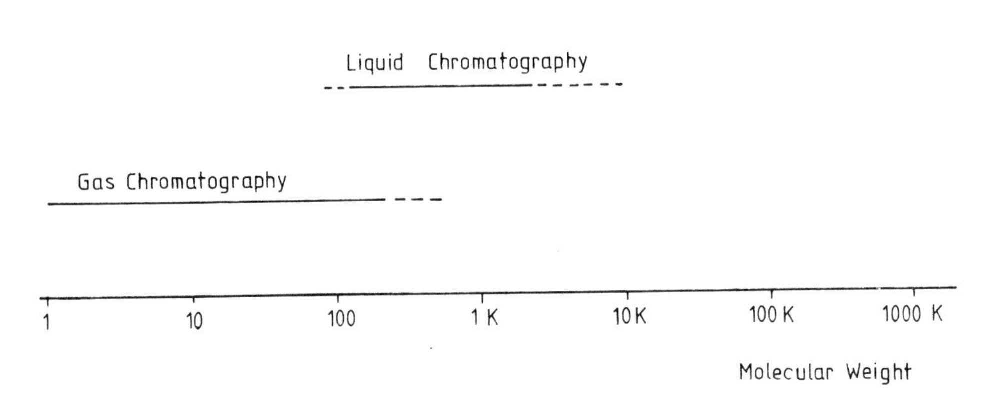

Fig.I.1. Scope of application of chromatographic techniques

rivatives that can be separated by gas chromatography. Consequently, the upper limit of molecular weights to which gas chromatography is applicable cannot be stated exactly.

Liquid-chromatographic methods are utilized chiefly for the separation of substances that decompose on vaporization. The LC separation of small and nonpolar molecules is difficult because their sorption on the stationary phase has to compete with that of the large excess of solvent molecules present. For larger molecules exclusion mechanisms become noticeable. Liquid chromatography makes a continuous transition into exclusion chromatography, where the separation depends solely on differences in molecular size. Figure I.1 shows schematically the range of applications for the three chromatographic methods.

Chromatographic separations can be accomplished by several methods. Based on the method of sample application used, one can distinguish the following [16]:

1. Continuous Sample Introduction Methods

The sample or its solution is continuously fed onto the column. This method is known as frontal analysis [17] or adsorptive filtration. If a solvent is used, it has to be selected to minimize its interaction with the stationary phase. Only the least retained substances can be isolated in pure form by this method, all the others appear as mixtures. This method can only be used discontinuously, for the column is exhausted when the mixture appears at the end of the column. It is therefore of little use for high pressure liquid chromatography, but is frequently employed for the purification of chromatographic solvents [5] and for the enrichment of trace compounds from liquids [21].

2. Discontinuous Sample Introduction Methods

a) Elution analysis. The sample is introduced into a continuous stream of eluent. The eluent composition before and immediately after sample introduction remains constant. If this persists throughout the entire separation and the interaction of the eluent with the stationary phase is small compared to that of the sample, one speaks of *elution analysis* or *elution chromatography* [18]. Typical elution chromatograms are thus obtained in which, ideally, each peak is separated from the following one by a band of pure eluent.

b) Gradient elution. If the interaction of the eluent with the stationary phase is increased during the analysis, i.e., the composition of the eluent is varied continuously and increases in elution strength, one speaks of *gradient elution* [19]. Gradient elution always reduces the analysis time. The more strongly retained sample components are eluted as much sharper, i.e., narrower, zones than in the case of elution analysis with constant eluent strength.

c) Displacement chromatography [20]. In this method the eluent is also changed after sample application to one whose interaction with the stationary phase exceeds that of all the sample components, so that it displaces them completely from the stationary phase. The displacer pushes all sample components in front of it in the order of increasing retention on the stationary phase. They appear in the column eluate in pure form, one behind the other, followed by the displacer. In contrast to elution chromatography, the individual components are not separated from one another by the pure eluent. The transition zones are mixed by diffusion, and therefore displacement chromatography has not found extensive application. Frequently, however, a displacer is added to complete a gradient elution. The column can be *regenerated* if the displacer can be washed out from the column by an eluent.

In separations, elution development is used almost exclusively because, in the ideal case, the peaks are separated by the pure eluent and consequently can be isolated in pure form. The position of the peak maximum for symmetrical peaks serves for qualitative identification of the sample components. The advantage of high pressure liquid chromatography over conventional chromatographic methods lies in the speed of analysis, and in the simple identification and quantitative determination of the separated components.

References Chapter I

1. Tswett, M.S.: Ber. dtsch. botan. Ges. *24*, 316, 384 (1906). Vgl. Hesse, G., Weil, H., in: Woelm-Mitteilungen Al 1, Eschwege 1954
2. Lederer, E., Lederer, M.: Chromatography 2nd ed. Amsterdam: Elsevier 1957
3. Lederer, E. (Ed.): Chromatographie en chimie organique et biologique. Vol. I und II. Paris: Masson 1959, 1960
4. Heftmann, E. (Ed.): Chromatography. New York: Reinhold 1969
5. Hesse, G.: Chromatographisches Praktikum. Frankfurt/Main: Akad. Verlagsges. 1968
6. Cramer, F.: Papierchromatographie. Weinheim: Verlag Chemie 1958
7. Hais, J., Macek, K.: Handbuch der Papierchromatographie. Bd. I. Jena: Gustav-Fischer-Verlag 1958
8. Stahl, E. (Ed.): Handbuch der Dünnschichtchromatographie, 2. Aufl. Berlin-Heidelberg-New York: Springer 1967
9. Randerrath, K.: Dünnschichtchromatographie. Weinheim: Verlag Chemie 1966
10. Giddings, J.C.: Dynamics of Chromatography. New York: Marcel Dekker 1965
11. Engelhardt, H., Weigand, N.: Anal. Chem. *45*, 1149 (1973)
12. Helfferich, F.: Ionenaustauscher. Weinheim: Verlag Chemie 1959
13. Dorfner, K.: Ionenaustauscher. Berlin: DeGruyter 1964
14. Determan, H.: Gelchromatographie. Berlin-Heidelberg-New York: Springer 1967
15. Altgelt, K.H., Segal, L. (Eds.): Gel Permeation Chromatography. New York: Marcel Dekker 1971
16. Halász, I.: Lecture notes, University of Nice, 1971
17. Tiselius, A.: Arkiv Kemi Min. Geol. *14b*, 22 (1941); cf. Endeavor *11*, 5 (1952)
18. Reichstein, T., van Euw, J.: Helv. Chim. Acta *21*, 1197 (1938)
19. Alm, R.S., Williams, R.J.P., Tiselius, A.: Acta Chem. Scand. *6*, 826 (1952)
20. Tiselius, A.: Arkiv Kemi Min. Geol. *16a*, 18 (1943)
21. Thesis; Aufsatz, M.: Saarbrücken, 1976

Chapter II

Fundamentals of Chromatography

A. Retention

The elution-chromatographic separation of two substances on a column can be likened to a steeple chase in that the time required for the substances to reach the end of the column depends on their retardation by the obstacles (the degree of retention by the stationary phase). The substances differ only in the time spent in or on the stationary phase; i.e., their net retention times t'_R are different. The total retention time t_R consists of this net retention time in the stationary phase and the time spent in the mobile phase t_o, also called the *dead time*.

$$t_R = t_o + t'_R \quad (1)$$

The dead time is the same for all substances; it is also the elution time of the solvent molecules.

The terms used in this chapter for the characterization of columns are explained in Figure II.1. The retention time *must* be independent of sample size if chromatography is to be employed for the qualitative identification of various substances. In other words, the ratio of the amount of sample in the stationary phase and in the eluent should be independent of the sample concentration. Only in such cases are symmetrical peaks obtained, which can be described by a Gaussian curve. The appearance of asymmetric peaks may indicate a nonlinear isotherm.

Because the retention time is dependent on the flow rate of the eluent, it is better to use the *retention volume*. The retention volume is the product of the retention time and the volume flow rate F (cm^3/min) of the eluent.

$$V_R = t_R \cdot F \quad (2)$$

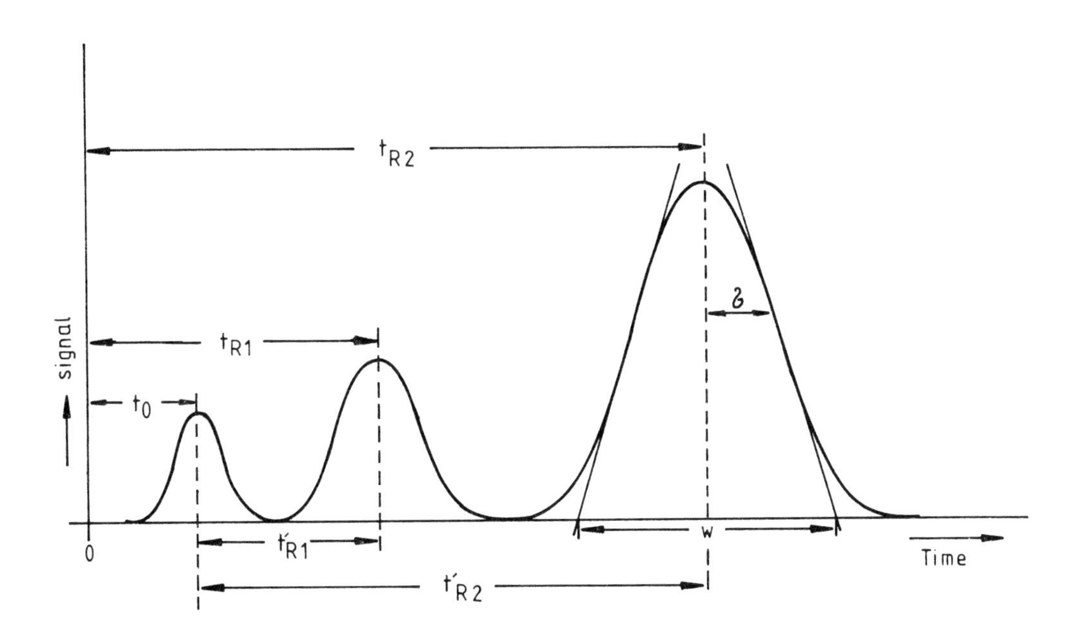

Fig.II.1. Illustration of important parameters for the characterization of separations. t_o is the dead time of a column, t_{R_1}, t_{R_2}, ... are the retention times of components 1, 2, ... whereas t'_{R_1}, t'_{R_2} are the corresponding net retention times. w is the baseline width - the distance between the intercepts of the tangents to the points of inflection - and equals 4σ. σ is the variance of the Gaussian curve. For further explanations see text

Similarly, the mobile phase volume V_M of the column can be determined from the dead time.

$$V_M = t_o \cdot F \quad (3)$$

The *net retention volume* V_N designates the total retention volume minus the mobile phase volume.

$$V_N = V_R - V_M \quad (4)$$

The net retention volume is proportional to the stationary phase volume V_S. The proportionality factor is the thermodynamic partition coefficient K, which is equal to the Nernst partition coefficient when the support has no measurable effect on retention (cf. Chapter VII).

$$V_N = K \cdot V_S \ . \tag{5}$$

In adsorption chromatography it is convenient to use a *normalized net retention* volume V_N^o, based on one gram of adsorbent:

$$V_N^o = V_N/g_A \tag{5a}$$

where g_A is the weight of adsorbent in the column.

Since the retention times can be taken directly from a chromatogramm, they are preferred, for reasons of convenience, to the more correct retention volumes. It should be stressed that when comparing retention times the volume flow rate of the eluent must be kept constant, and when comparing retention volumes the free cross-section of the column must be held constant. Because the latter in particular cannot be determined exactly, the linear flow rate u is used in chromatography.

The ratio of the times spent in the stationary and in the mobile phases is called the capacity factor or *mass distribution ratio* k'.

$$k' = \frac{t'_R}{t_o} = \frac{t_R - t_o}{t_o} \ . \tag{6}$$

The k'value is independent of column length and in LC is not a function of the mobile phase velocity, provided it is less than 5 cm/sec. At equilibrium the following relationship exists between the k' value and the partition coefficient K:

$$k' = K \cdot \frac{V_S}{V_M} \ . \tag{7}$$

In chromatography the k' value is preferred over the partition coefficient because accurate determination of the phase ratio V_M/V_S is difficult. The capacity ratio k' is independent of the sample size as long as one operates in the linear region of the isotherm and maintains all external conditions constant. These are essentials for reproducible chromatographic separations.

The ratio of two capacity factors for the same stationary phase obtained under constant external conditions is designated as the *relative retention* α:

$$\alpha = \frac{t_{R_2} - t_o}{t_{R_1} - t_o} = \frac{t'_{R_2}}{t'_{R_1}} = \frac{k'_2}{k'_1} = \frac{K_2}{K_1} \,. \tag{8}$$

The relative retention is a measure of the selectivity of the separation system, and by convention is always greater than unity. The more selectively a stationary phase retains one of two components, the greater is the relative retention of these components. If $\alpha = 1$, there are no thermodynamic differences between the two components in a given system, and they cannot be separated. The effect of the magnitude of the relative retention on resolution will be discussed in detail later.

Since in chromatography equilibrium is almost always attained [1,2], the relative retention α is a thermodynamic quantity which, at constant temperature, depends only on the nature of the sample and the properties of the stationary and mobile phases. In HPLC the α values may change during the lifetime of a column because of the stripping away of the stationary phase or the coating of the support by impurities in the eluent, whose presence is not always obvious. The determination of the relative retentions for various pairs of substances on the same column during its lifetime provides a useful means of detecting gradual changes in the properties of the column. This test should be repeated frequently.

B. Linear Flow Rate, Porosity, Permeability

The velocity of the eluent is preferably specified as the average linear velocity u (cm/sec) rather than the volume flow rate F (cm^3/sec). It corresponds to the average velocity at which solvent molecules move through a column. The linear flow velocity is independent of the column cross-section and is proportional to the pressure drop along the column. It can be calculated with the aid of the dead time and column length L:

$$u = L/t_o \,. \tag{9}$$

The linear flow rate can also be calculated from the volume flow rate F and the free cross-section q of the column.

$$u = F/q = \frac{F}{r^2 \pi \varepsilon_T} . \tag{10}$$

The free cross-section q of a packed column is always a fraction of the actual cross-section of the column, and depends on the type of packing. In a column packed with glass beads only about 40% of the cross-section of the empty column is available to the eluent. This applies only to regularly packed columns, i.e., those whose internal diameter is greater than about 10 times the average particle diameter. The fraction of the cross-section of the unfilled column that is available to the eluent, is called the total porosity ε_T. Accordingly, for nonporous glass beads $\varepsilon_T = 0.4$ [4,5].

The porosity ε_T of a chromatographic column can be calculated from the volume flow rate F and the linear flow rate u [5,6]:

$$\varepsilon_T = \frac{F}{u \cdot \pi r^2} = \frac{F t_o}{L \cdot \pi r^2} = \frac{F \cdot t_o}{V_o} \tag{11}$$

where V_o is the volume of the empty column.

For columns packed with porous materials a larger porosity is always found. Silica with a pore volume of about 1 ml/g has a porosity of about 0.85. This means that the free cross-section of columns packed with these porous materials is about double that of columns packed with nonporous particles. Since the volume between the particles is practically independent of whether the particles are porous or nonporous, virtually the same volume must be inside the particles. Distinction is therefore made between a "stagnant" mobile phase in the pores and a "moving" mobile phase. For totally porous particles such as silica these two volumes are about equal. Consequently, the average linear flow rate in columns regularly packed with nonporous particles is about double that of those containing porous particles (at constant volume flow rate and for the same pressure drop). Because in chromatography, by definition, there is no transport within the pores of the stationary phase, the eluent velocity between (outside) the particles is the same, and is termed the interstitial velocity.

If the dead time and linear velocity of columns packed with porous materials are determined with (unretained) sample molecules of various sizes, different values are obtained, depending on the portion of the pore volume that is accessible to the molecules. Evidently, molecules of different sizes migrate at different rates through columns containing porous supports, the larger moving more rapidly than the smaller. The range of linear velocities (migration rates) lies between that obtained with high-molecular-weight molecules that are totally excluded from the pores (the interstitial velocity) and that of the eluent, measured in the usual chromatographic manner. However, with molecules smaller than those of the eluent, e.g., N_2, H_2, He, even slower migration rates may be obtained. Nevertheless, the average linear velocity always refers to that of the eluent. Separations by exclusion chromatography (cf. Chapter IX) are based on such differences in the migration rates of the samples.

The linear velocity obtained is proportional to the pressure drop along the column. The relationship is given by the permeability K_F (cm^2) of the column.

$$K_F = \frac{F \cdot \eta \cdot L}{\Delta p \pi r^2} = \frac{u \cdot \eta \cdot L \cdot \varepsilon_T}{\Delta p} = \frac{L^2 \eta \cdot \varepsilon_T}{\Delta p \cdot t_o} \qquad (12)$$

where F is the volume flow rate (cm^3/sec), u is the linear flow rate (Eq.(9)) in cm/sec, L - column length (cm), r - column radius (cm), η - eluent viscosity (Poise = 0.1 Pa·s) and Δp - the pressure drop (Pa ≈ 10^{-5} atm).

The relationship between permeability and particle size in chromatographic practice is given to a good approximation by the following rule of thumb [6]:

$$K_F = d_p^2/1000 \qquad (13)$$

where d_p represents the mean particle diameter in cm. A comparision of Eqs.(12) and (13) shows that for a given column length at constant linear velocity the required pressure drop is inversely proportional to the square of the particle diameter. When particles with a diameter < 10 μm are used, the upper pressure limit (300-400 atm) is reached at relatively low linear velocities (~1 cm/sec) with 30 cm columns. It should also be pointed out that the porosity and permeability have been defined in a way that is most expedient for chromatography. These definitions differ from those used in hydro- and aerodynamics [5].

When the average particle diameter is known, the permeability can be estimated from Eq.(13) and than compared with the measured value based on Eq.(12). The efficiency of a column packing can be assessed from this value. However, the reverse aspect is much more important because the particle size distribution is very difficult to determine for particle diameters of < 10 μm. It has therefore been proposed [7] that the average "hydrodynamic" particle diameter for packed columns be defined on the basis of Eq.(13). It has been shown that for columns wet-packed with silica gel the particle diameters determined in this way agreed well with the actual number-averaged diameter determined photometrically. For dry-packed columns somewhat poorer permeabilities were generally obtained [6,8].

C. Band Broadening

A sample band passing through a column is broadened or spread by diffusion processes. In principle, the considerations employed in gas chromatography [9,10] can be readily applied to liquid chromatography if the quantitative differences between the properties of gases and liquids are taken into account [11] (cf. Table II.1).

Table II.1. Order of magnitude values of parameters important in band broadening

	Gas	Liquid
Diffusion coefficient D [cm^2/sec]	10^{-1}	10^{-5}
Density ρ [g/cm^3]	10^{-3}	1
Viscosity η [Poise = 0.1 Pa·sec]	10^{-4}	10^{-2}
Reynold's number	10	100

Thus, the diffusion coefficients in liquids are about 10^4 times smaller than in gases. The viscosities of liquid mobile phases are greater by about a factor of 100 than those of gases. Whereas the interactions between mobile and stationary phases are negligible in GC, they play an essential role in LC. The theoretical treatment of LC is certainly simpler than that of GC because the liquid mobile phases are not compressible over the pressure range used.

The *plate height* H (height equivalent of a theoretical plate) serves as a measure of band broadening. The plate height in chromatography, in contrast to that in distillation, is defined for a single component at a given eluent velocity and constant phase ratio and temperature.

Equation (14) indicates how the plate height can be determined from a chromatogram.

$$H = \frac{L}{16} \cdot \left(\frac{w}{t_R} \right)^2 . \tag{14}$$

The peak width w is measured as the distance between the intersections of the baseline with the tangents to the points of inflection. For a Gaussian curve this distance is 4σ. Analogous to distillation, one also uses the number of theoretical plates N.

$$N = \frac{L}{H} = 16 \left(\frac{t_R}{w} \right)^2 . \tag{15}$$

The plate number, in contrast to plate height, is proportional to column length. But even a column having 10 000 plates cannot separate two components that have the same k' values. It is therefore better to use the *effective plate height* (16) or *effective plate number* (17)[12]:

$$H_{eff} = H \cdot \frac{(1 + k')^2}{k'^2} = \frac{L}{16} \left(\frac{w}{t'_R} \right)^2 . \tag{16}$$

$$N_{eff} = N \cdot \frac{k'^2}{(1 + k')^2} = 16 \left(\frac{t'_R}{w} \right)^2 . \tag{17}$$

The effective plate height or number is a constant for each sample if column conditions are held constant. Since these effective quantities are proportional to the resolution R of two components (cf.II.D), they represent a measure of the separation efficiency of a chromatographic column.

D. Resolution

The resolution R of two sample bands is defined in terms of the distance between the two peak maxima (expressed as the difference in the two retention times) and the arithmetic mean of the two band widths w.

$$R = \frac{2(t_{R_2} - t_{R_1})}{w_1 + w_2} \quad . \tag{18}$$

In chromatography one strives not for greater, but rather *optimum resolution,* i.e., the peaks should be separated from each other only as far as necessary. If Gaussian curves are obtained, a resolution of 1.5 (also called 6σ separation because $w = 4\sigma$ for Gaussian curves) suffices for quantitative analysis. In this case the peaks are separated almost to the baseline from each other. However, greater resolution is achieved at the expense of analysis time. For a resolution of 1.0 the distance between peak maxima is equal to w or 4σ (hence the term 4σ separation), which is still adequate for quantitative analysis, since there is only about 2% peak overlap. Equation (18) can be related to the other chromatographic parameters [13] as follows:

$$R = \frac{1}{4}\,\frac{\alpha - 1}{\alpha}\sqrt{N_{eff,2}} = \frac{1}{4}\,\frac{\alpha - 1}{\alpha} \cdot \frac{k'_2}{1 + k'_2}\sqrt{N_2} \quad . \tag{19}$$

This is the most important equation of chromatography because it combines the factors on which a separation is based (relative retention α, mass distribution ratio k') with the factor that counteracts separation, the band broadening (plate number N).

By rearranging Eq.(19) we get

$$N_{eff} = \left(\frac{4R\alpha}{\alpha - 1}\right)^2 \tag{20}$$

from which the number of plates required for a separation can be calculated for a given α and the desired resolution R. Table II.2 presents some values calculated by means of Eq.(20). A resolution of 1.5 always requires more than double the number of effective plates needed for a resolution of 1.0. It is also evident that the number of required plates rises rapidly as α approaches small values, especially

Table II.2. The number of plates required to achieve a desired resolution for a given α

Relative retention α	R = 1.0	R = 1.5
1.005	650 000	1 450 000
1.01	163 000	367 000
1.02	42 000	94 000
1.05	7 100	16 000
1.07	3 700	8 400
1.10	1 900	4 400
1.15	940	2 100
1.25	400	900
1.50	140	320
2.0	65	145

as α becomes < 1.1, i.e., as the properties of the substances to be separated become very similar. If the column efficiency is inadequate for a given α, a more selective system must be sought for this pair. In HPLC this can be achieved by varying the stationary and/or the mobile phase.

Figure II.2 should clarify the interaction of both of these factors on the resolution of two peaks. If the relative retention of two components is large, satisfactory separation can be obtained even if the peaks are very broad, i.e., when column efficiency is low (A). In this case the selectivity of the column governs the separation.

For smaller relative retentions no useful separation is attained with the same column (B). By increasing the column efficiency, however, components with such small relative retentions can be separated (D). With large relative retention and good column efficiency a resolution that considerably exceeds the optimum (R = 1.5) is obtained (C). Since the plate number is proportional to column length, the column can be shortened, thereby reducing the plate number and saving analysis time.

A far greater improvement in the resolution of two peaks can be achieved by changing the separation conditions rather than by increasing the plate number. Doubling of the plate number by doubling column length would enhance resolution only by a factor of 1.4. Simultaneously, the retention and analysis times would be doubled. In practice,

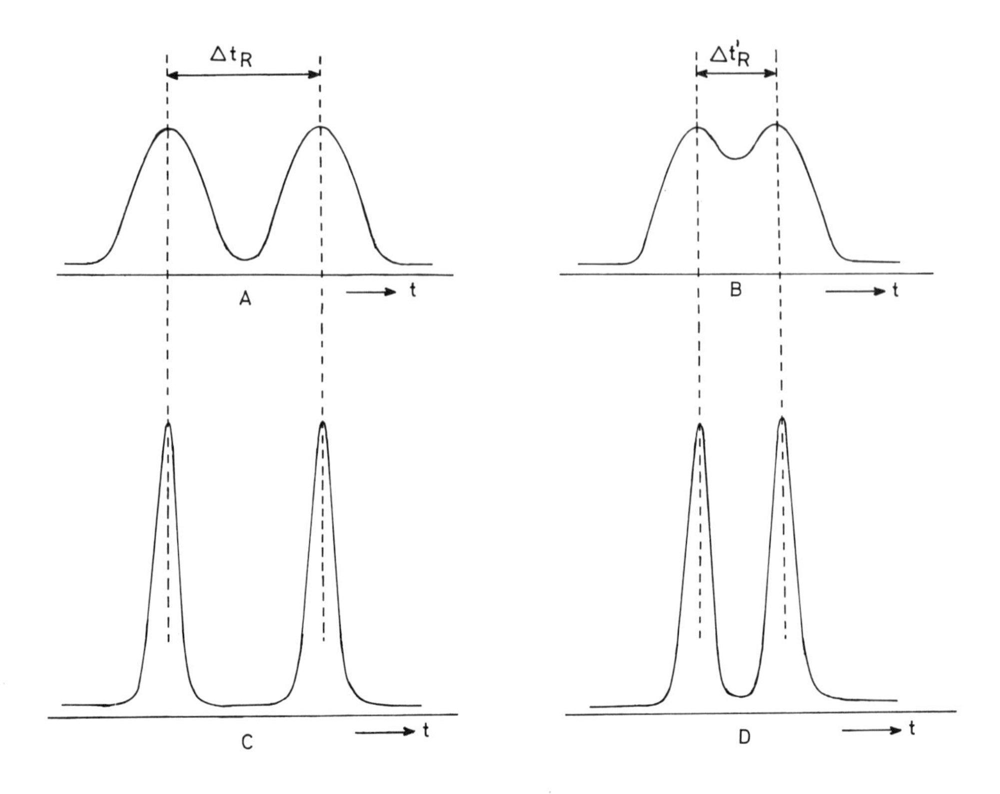

Fig.II.2. Illustration of the effect of relative retention and efficiency on resolution (for explanation see text)

however, high resolution can frequently be achieved more conveniently by increasing the plate number than by altering the separation conditions. Therefore, in chromatography (especially for theoretical considerations) the separation efficiency of a column should always be optimized.

The larger the difference in the partition coefficient, the greater is the resolution of two peaks because of the increased relative retentions. Since the number of plates required for a particular separation is proportional to the analysis time, a larger α implies a shorter analysis time. The separation system should therefore be selected so that the relative retentions are as large as possible, i.e., that the system is very selective.

E. Dependence of Band Broadening on Flow Rate

Experimental measurement of the dependence of band broadening on flow rate in gas or liquid chromatography leads to the well-known curve depicted in Fig.II.3. This curve can best be described by the *van Deemter* equation, which has general validity for chromatography [14].

$$H = A + \frac{B}{u} + C_M \cdot u + C_S \cdot u \ . \qquad (21)$$

Detailed discussion may be found in monographs on gas chromatography, e.g.[9,10].

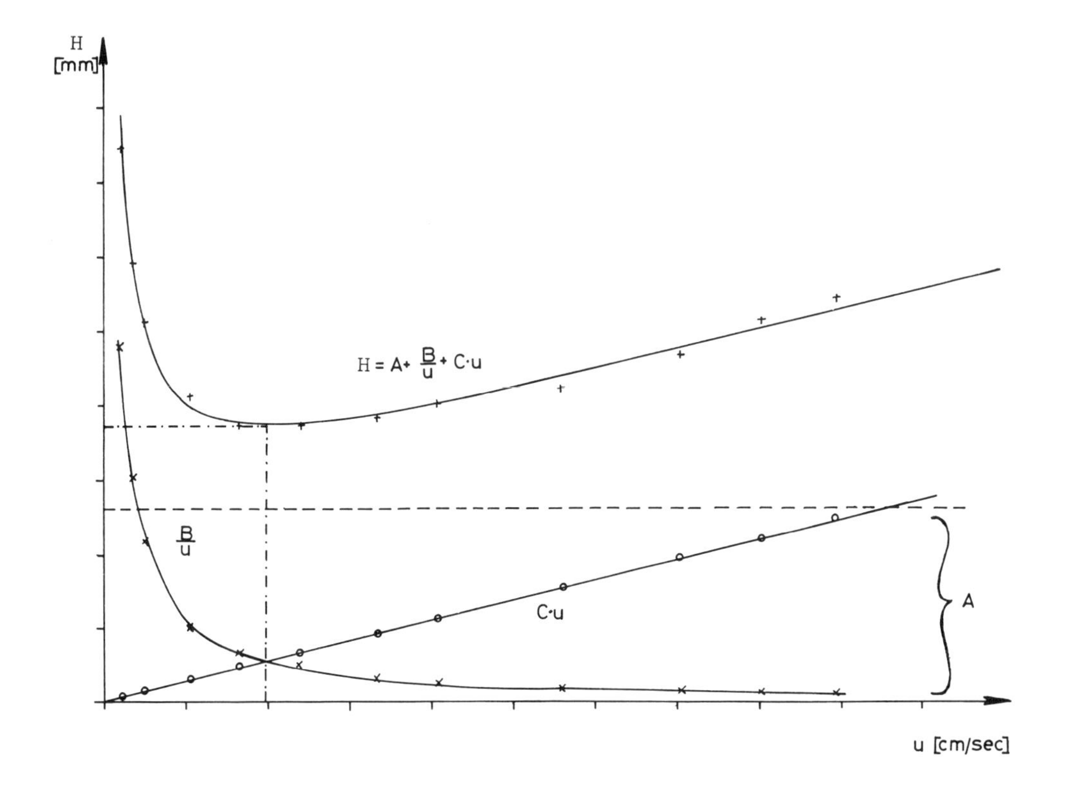

Fig.II.3. Dependence of band broadening on flow rate (schematic)

The contribution to band broadening that is independent of the flow rate has been attributed to so-called eddy diffusion (A term). When a sample band migrates through a packed column, the individual flow paths around the packing particles are of different lengths. These variations in the flow direction and rate lead to band broadening that should only depend on the efficiency of the column packing. The A term is proportional to the particle size and is usually given as follows:

$$A = 2\lambda d_p \tag{22}$$

where λ represents the so-called "packing factor", whose numerical value varies between 1 and 2.

The B term of Eq.(21) is designated as the longitudinal diffusion term and becomes noticeable only at low flow rates. In LC its contribution to band spreading can very often be neglected at flow velocities > 0.5 cm/sec, particularly if d_p is > 10 μm. In the literature this diffusion effect in the eluent is described as

$$B = 2\gamma D_m \tag{23}$$

where γ is a constant that takes into account the restricted diffusion path in the packed column.

The diffusion of the sample molecules, including those of the unretained substance, is different in the moving eluent as compared to the stagnant one. In LC the effective diffusion in the axial direction is greater at higher flow rates than the longitudinal diffusion (cf. the Taylor equation). This diffusion contribution in the moving eluent is contained in the C_m term. It is a function of the particle diameter and is inversely proportional to the diffusion coefficient in the mobile phase.

$$C_m = \phi \frac{d_p^2}{D_m} \tag{24}$$

where ϕ is only a function of the capacity factor k'.

During the migration through the column, the sample molecules are constantly passing from the mobile to the stationary phase (sorption) or vice versa (desorption). When a molecule is sorbed, it is left behind the band center which continues to migrate down the column. When it passes out of the stationary phase back into the mobile phase, it migrates faster than the center of the retarded band because the eluent velocity is always greater than the average migration rate of a sample band. The so-called *mass transfer term* C_S in the stationary phase is given as follows:

$$C_S = \text{const} \cdot f(k')\frac{d_f^2}{D_S} \quad . \tag{25}$$

The fraction d_f^2/D_S has the dimensions of seconds and represents a measure of the residence time in the stationary phase. In the GC literature d_f is designated as the average film thickness of the stationary phase, and D_S is the diffusion coefficient of the sample in the stationary phase.

Because the rates of diffusion in the mobile and stationary phases are of the same magnitude in LC, the diffusion within the stagnant mobile phase in the pores also contributes to band broadening, in contrast to that in GC. Molecules that diffuse into this stagnant mobile phase are *retarded* relative to those that remain in the moving phase. This contribution to band spreading, which has nothing to do with sample retention, affects both unretained and retained substances. The effect is relatively small for slow migration rates (high k' values). Its contribution to band broadening can be reduced by decreasing the pore depth, thus shortening the diffusion distance. This can be achieved in two different ways: either by utilizing *porous layer beads* (PLB)[15] in which case a solid core, such as a glass bead, is coated with a thin (~1 μm) porous layer, or by decreasing the particle diameter and thereby, of course, also the pore depth.

The PLB have very small pore volumes, resulting in very short diffusion distances in and out of the pores. More rapid mass transfer and smaller band broadening is obtained with these than with totally porous particles of the same diameter. Their disadvantage lies in their low capacity, as the total amount of stationary phase in a given column is small. They have been employed with considerable success for achieving rapid LC separations.

The importance of the PLB diminished as it became possible to pack efficient columns easily and reproducibly using particles with diameters around or less than 10 μm. In contrast to PLB columns the load capacity of those packed with such small particles was substantially greater and the k' values were higher as a result of the larger amount of active stationary phase on the column.

PLB can be used to advantage when excessively long retention times are obtained on totally porous particles with a given eluent. Because of the smaller amount of stationary phase, the retention times are substantially shorter on PLB.

All of these considerations should be regarded to be more or less qualitative because the processes occurring in a packed column cannot be described exactly. Exact treatment is possible only for open tubes whose inner wall is covered with a thin film of stationary phase (Golay equation).

In additional to the van Deemter equation there are also simpler equations in the literature that relate band spreading to flow rate. For a certain velocity range, the dependence of band broadening on flow velocity can be described by the empirical approximation

$$H = A^{*} + C^{*} \cdot u \tag{26}$$

where A^{*} is the flow-independent contribution to band spreading, but beyond that has no physical significance. The slope of the curve (the C^{*} term) gives an indication of the rate of mass transfer.

Another empirical approximation for relating band broadening to flow velocity was introduced by Snyder [17]:

$$H = D \cdot u^{x} . \tag{27}$$

Subsequent investigations revealed that x is not always 0.4 as assumed originally, but varies over the range of 0.3 to 0.6, depending on the column packing. The constant D is the plate height obtained at a linear flow rate of 1 cm/sec and contains both the A and C terms of Eq.(26).

Using the reduced, dimensionless parameters introduced by Giddings [9], such as reduced plate height $h = \frac{H}{d_p}$ and reduced flow rate $\nu = \frac{ud_p}{D_m}$, Knox presented another semi-empirical equation to describe the variation of band spreading with flow velocity [25,26,27]:

$$h = \frac{B}{\nu} + A \cdot \nu^{0.33} + C \cdot \nu \tag{28}$$

where B represents the axial diffusion, which usually lies between 1.5 and 2, and C the restricted mass transfer in the stationary phase, which typically assumes values of 0.01 to 0.02. The A term is associated with the quality of the column packing and is supposed to describe the restricted mass transfer in the mobile phase outside the particles. Such curves are apparently independent of the properties of the mobile phase and the particle diameter. Usually, minima are obtained at h values of from 2 to 5 and at the corresponding reduced velocities between 2 and 10. It was shown [28,29] that by optimizing the particle diameter and the column cross-section, one can operate at a given pressure drop under the optimum conditions described here.

Based on the interpretations presented here, the lowest attainable H value under optimal conditions would always be more than twice the average particle diameter at the optimal linear velocity of less than 1 mm/sec (if a typical value of ca. $3 \cdot 10^{-5}$ cm/sec is assumed for the diffusion coefficient in heptane or methylene chloride).

F. Band Broadening and Particle Size

The variation of the plate height with the square of the particle diameter (Eq.(24)) and other similar proportional relationships assumed in certain empirical approximations (e.g., Eq.(28)) have been verified experimentally. However, this appears to be strictly valid only down to 80 μm diameter particles [8]. In HPLC it was found that in going to smaller and smaller particle sizes the decrease in the plate height no longer corresponds to this proportionality, but is smaller. The values in the literature vary between $d_p^{1.3}$ and $d_p^{1.8}$ [7,16,17,24,30,31].

In determining this proportionality a question arises concerning the average particle size of a sieve fraction. The particle diameter d_p is not an unambiguously defined quantity, but is rather a function of the way it is determined (e.g., light scattering, microscopic measurement, sedimentation, Coulter Counter, etc.) and the method by which it is averaged (number, volume, or weight averaged). Moreover, possible changes during column packing in the average diameter of the particles or in their distribution are usually disregarded. Halász proposed [7] determining the effective "hydrodynamic" particle diameter of a packed column in terms of its permeability, i.e., in terms of the pressure required to achieve a desired flow rate, and to define the "average particle diameter" in this manner. By rearranging Eq.(13) one obtains

$$d_p = \sqrt{1000 \cdot K_F} = \sqrt{\frac{1000 \cdot F \cdot \eta \cdot L}{\pi r^2 \cdot \Delta p}} \quad . \qquad (29)$$

For narrow sieve fractions the particle diameter defined by Eq.(29) agreed with the number-average diameters determined microscopically.

Examination of the dependence of the individual terms of the van Deemter equation (Eqs.(21-25)) on particle size reveals that both the B term (Eq.(23)) and the mass transfer term in the stationary phase C_s (Eq.(25)) are independent of the particle diameter. In LC the latter term is negligible in comparison to the mass transfer term, C_m, in the mobile phase (Eq.(24)), provided that "heavily loaded" columns are not used.

From experimental results it can be shown that $A \sim 2 - 3\ d_p$ (i.e., $\lambda \sim 1 - 1.5$) and $B = 2\ D_m$ (i.e., $\gamma = 1$). The C_m term (Eq.24)) contains the ϕ function that is dependent on the k' value. If it is assumed, as does Halász [24], that the Golay term [33] for mass transfer in the mobile phase is valid, a value of 0.047 is calculated for ϕ when k' = 1. (For k' = 5 a value of $\phi = 0.09$ is obtained. Since the H values are only slightly dependent on the k' values for $d_p < 10\ \mu m$, the assumption that $\phi = const = 0.047$ appears to be justified.

Under these assumptions Eq.(21) becomes

$$H = 3\ d_p + \frac{2D_m}{u} + \frac{0.047 \cdot d_p^2}{D_m} \cdot u\ . \qquad (30)$$

In calculating the H values, all terms must be included because a large A term can compensate for a small C term, and vice versa. This equation holds for the typical velocity range, which is only limited by currently available equipment.

In considering only the order of magnitude dependence of the individual terms on d_p, it becomes evident that A decreases linearly with d_p but C with d_p^2. In going to small particles ($d_p < 10\ \mu m$) the proportionality between band broadening and d_p decreases as the particle size becomes smaller because $A \gg (B/u + C \cdot u)$. For very small particles ($d_p < 3\ \mu m$) H is proportional to d_p [24]. Consequently, with decreasing d_p the gain in separation efficiency diminishes continuously at the expense of the required pressure because the permeability is always proportional to d_p^2.

In order to achieve a high resolution, one should always operate at u_{min}, the velocity corresponding to the minimum of the H values. (Of course, to reduce the analysis time one can work at $u > u_{min}$. However, the efficiency would then diminish). The position of u_{min} is also a function of the particle size and is shifted to higher velocities with decreasing d_p. This means that with decreasing particle size additional pressure must be expended just to carry out the separation at u_{min}.

By differentiating Eq.(30) with respect to u, u_{min} can be calculated as

$$u_{min} = \sqrt{\frac{B}{C}} = \sqrt{\frac{2\ D_m^2}{o\ d_p^2}} \sim \frac{6.52 \cdot D_m}{d_p} . \tag{31}$$

From Eqs.(30) and (31) the minimum H value can be calculated as

$$H_{min} \sim 2\text{-}4\ d_p . \tag{32}$$

For small particle sizes, optimum H values are obtained at the minimum of about 2 - 4 particle diameters [34]. A reduction in the particle size (d_p < 5 µm) by a factor of 2 leads also to a two-fold decrease in H, whereas the required pressure rises by a factor of 8. The latter is the result of a decrease in the permeability (with d_p^2) and the two-fold shift in the u_{min} value to higher flow rates, which requires a higher pressure in order to operate at least at u_{min} while keeping the other parameters (e.g., column length, viscosity, etc.) constant. Since the equipment limits the pressure drop attainable, it, among other things, also determines the smallest particle size for routine HPLC work.

Moreover, still other difficulties stand in the way of using very small particles. For a given pressure limit imposed by the equipment, shorter columns must be used as the particle size is reduced in order to at least attain u_{min}. A shortening of the column to less than 7 - 10 cm is not recommended because the dead volume outside the column would no longer be negligible in comparision to the mobile phase volume inside the column. The extra-column band spreading (cf. Section G) then attains the same order of magnitude (or becomes even greater) than that inside the column. Furthermore, the peak widths measured in time units become so small that substantially greater demands than usual are placed on the response times of the detector and recorder.

One of the principal difficulties is the heat of friction caused by the pressure drop [24], which raises the eluent temperature of a completely insulated column by 5 - 7°C per 100 atm of pressure drop between the inlet and outlet. Hence, an axial temperature gradient is formed. Since a column is never an adiabatic system, an additional radial temperature gradient appears. As a result, the eluent viscosity, the interdiffusion coefficient of the sample and its retention behavior (partition coefficient) vary from place to place inside the column. In addition to the theoretical consequences (no exact description of the transport processes in the column is possible), this also leads to possible additional distortion of the sample bands (increased plate height).

For these reasons, the particle size cannot be decreased indefinitely in LC. At present, the limit appears to lie at 3 μm, and the optimum values apparently fall between 3 - 5 μm. At the current level of technology, a great deal of experience is required to pack good columns consistently with particles in this size range. Furthermore, the column length should exceed 7 cm when such particle sizes are used.

Packing columns with 10 μm particles is far easier. The separation efficiency of such columns, which are usually 20 - 30 cm long, is adequate for many routine purposes and can be increased by connecting several columns in series because they still have good permeability.

G. Extra-Column Band Broadening

As a consequence of the low diffusion coefficients of sample molecules in the liquid mobile phase, considerable contribution to band spreading results from the dead volume in the injection block, in the connection between the column and the detector, and from the detector volume itself. The greater the volume of the connections and the detector cell in relation to the sample retention volume, the more troublesome is the band broadening [19]. Due to the slow diffusion, the developing stream profiles are not equalized. This can go so far that the peak of a pure substance exhibits two peaks and thereby gives the false impression of the presence of an impurity or a second component. This type of band broadening becomes disturbingly evident for substances with small k'values that migrate rapidly through the column. The band spreading in the connecting tubing can be reduced or even completely eliminated by producing artificial radial mixing [20]. Flattened, twisted, or coiled capillary tubing in which the streampaths are frequently forced to change direction should be used when long connections cannot be avoided or a *heat exchanger* is required before the detector [21]. In sample application the dead volume can be relatively easily avoided if care is taken to inject the sample on the top of the column. The detector cell volume should be as small as possible, within the requirement that the optimum response requires a certain path length. A detailed discussion and theoretical treatment of these problems can be found in [22].

The effect of band spreading outside the column can best be established by measuring H *vs* u curves for several substances. If the curves with low k'values ($k' < 3$) exhibit a different shape from those with higher k'values, the presence of extra-column band broadening is indicated. Also, the peak form, e.g., tailing for substances with small k'values and its absence for the more strongly retained ones, points to such effects. A greater plate height for the unretained peak than for a retarded one suggests extra-column band broadening. Home-made and commercial instruments, particularly after modification, should always be checked for such effects.

H. Optimum Analysis Conditions and Analysis Time

A separation is optimum if it is as complete as necessary in the shortest time [23]. The technical difficulties, such as the operational temperature and pressure, should be small. As mentioned previously, a resolution $R > 1.5$ is undesirable because it can only be achieved at the expense of analysis time. For every mixture there is an optimum system of stationary and mobile phases with an optimum temperature that posses the greatest selectivity for the desired separation and thus provides a large relative retention (α). One should always strive for such a system because it requires the smallest number of plates and a shorter column for the separation.

The analysis time required under these conditions, which corresponds to the retention time of the last peak, is given [23] by:

$$t_{req} = H_{eff} \cdot N_{eff} \cdot \frac{L \cdot \eta}{k \cdot \Delta p} (1 + k'_2) \ . \qquad (33)$$

The analysis time required for an optimum separation depends on the mobile phase properties and the flow rate attained with a given pressure drop. Therefore, when a choice exists between several *eluents* with similar chromatographic properties, the one with the lowest *viscosity* should be selected.

The column properties and packing exert considerable influence on the analysis time. As stated, columns should be as short as possible. However, a certain length is necessary to attain the required plate number. Since the plate number is inversely proportional to the square (or a smaller power) of the particle size, smaller particles should always be used. With small particles, however, the *permeability* of a column is diminished, i.e., to maintain a constant eluent flow rate requires the application of a higher pressure. The usual high pressure pumps restrict the available pressure drop to 300 to 400 atm, so that instrumental limitations stand in the way of reducing the particle size and, hence, the analysis time. Moreover, a fundamental limitation (the position of the minimum of an H *vs* u curve) appears to make it impractical to work with particle sizes of less than 3 - 4 μm (cf. Section F).

The temperature affects the analysis time only indirectly. The viscosity of the mobile phase decreases with rising temperature, resulting in increasing eluent velocities for a constant pressure drop. Since the rate of diffusion also increases with rising temperature, sharper peaks are obtained at higher temperatures. Of course, the kinetics and thermodynamics of sample retention are also influenced by temperature.

For a given separation the *required plate number* can be obtained with a long column packed with larger particles (~ 30 μm) as well as with a shorter one containing smaller particles (5 or 10 μm). It is assumed, however, that both columns can be packed equally well. Which column is used depends on the instrumental capabilities, i.e., if the pressure necessary for a constant flow rate can be attained. The stability of the column and packing to pressure plays a role as does the *instrumental dead volume* which presents a substantially greater problem for short columns. However, the speed of analysis is always greater with shorter columns packed with smaller particles. Another advantage of small particles lies in the higher detection sensitivity they permit, which results from the sample components migrating as sharper zones, hence reaching the detector less diluted and giving rise to higher peaks (cf. Section XI.C).

A measure of the speed of analysis is the number of plates generated per second. Since the resolution of two peaks is proportional to the effective number of plates, it is advantageous to use the *effective plates* generated per second as a measure of the speed of analysis:

$$\frac{N_{eff}}{t} = \frac{u}{H} \frac{k'^2}{(1 + k')^3} \quad . \tag{34}$$

The greater the number of effective plates per second, the shorter is the analysis time. With very good columns packed with 5 - 10 μm particles 100 - 300 plates per second can be readily achieved. This corresponds to about 10 - 40 effective plates, depending on the k'value. The term containing the k' values in Eq.(34) has a maximum at k' = 2. For *rapid analyses* the system should be selected so that the k' values fall into the range of 1.5 to 4. *Multicomponent analyses* , of course, cannot be carried out in this narrow k' range. For them, the separation efficiency must be increased at the expense of the analysis time, by either operating at a slower flow velocity or employing a longer column.

For every column there is an *optimum flow rate* above which an increase in pressure fails to reduce the analysis time. The speed of analysis is proportional to H/u, whereas the eluent velocity is proportional to the applied pressure. A plot of H/u *vs* u and the H *vs* u curve from which it was derived are shown in Fig.II.4 (for convenience the plate height H was used, which is proportional to the effective plate height H_{eff}). The optimum flow rate is defined to occur where the H/u *vs* u plot approaches the horizontal. A substantial increase in pressure and, hence, in the linear velocity would produce no appreciable increase in the speed of analysis. Only about two thirds of the maximum possible speed of analysis is obtained at the point (~6 mm/sec) where the curve shown in Fig.II.4 approaches the horizontal. Therefore, in LC a further increase in the flow rate - if sufficient pressure is available - will always increase the speed of analysis. One should always strive to keep the slope (C term) of the H *vs* u curve as small as possible to minimize the decrease in the plate number as the flow rate is increased. Assuming the A term is small, the smaller the C term the lower is the optimum velocity. On the other hand, if the A term is large and the C term small, the optimum flow rate will be high. From the discussion of the optimum separation parameters it follows that for a good column not only should the band broadening be minimized, but the ratio of the A and C terms should be balanced out to obtain the smallest possible H/u values even at low flow velocities.

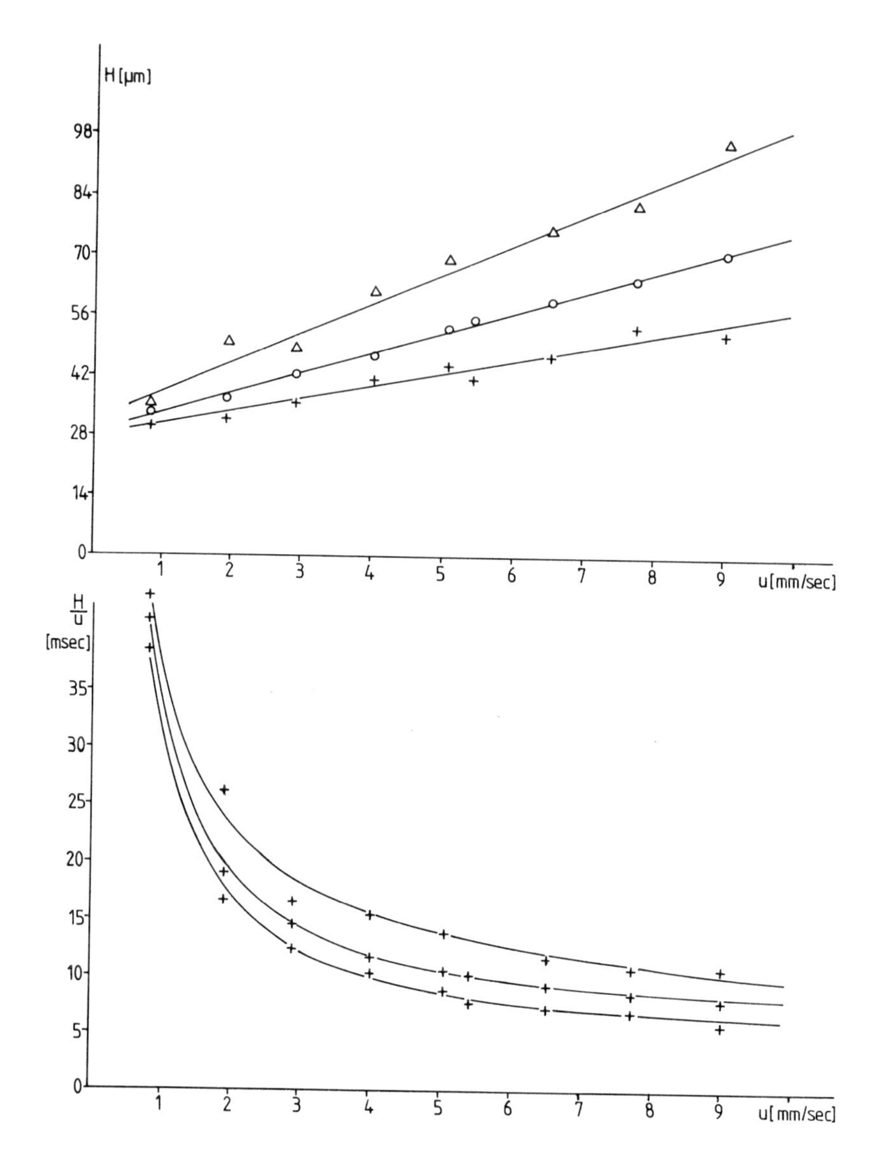

Fig.II.4. Band broadening as a function of flow rate (upper curve) and the determination of the optimal velocity (Stationary phase: silica $d_p \sim 10$ µm; eluent: n-heptane; column: 4 mm i.d., 30 cm long)

It should be pointed out again that the plate number is not the decisive factor for a separation: Simultaneously, there must be a sufficiently large difference in the k' values.

I. Selection of a Suitable Column

The preceding discussion suggests that a column may be selected on the basis of three criteria, namely

1) the attainable *resolution*,
2) the speed of analysis, and
3) the load capacity of the column.

If rapid analysis with good *resolution* is desired, it is expedient to use a column packed with PLB (porous layer beads). However, the amount of active stationary phase in the column is very limited in this case, and the system becomes overloaded with relatively small amounts of sample. High resolution can be achieved most readily by using the minimum sample size and a long analysis time. Sufficient pressure should be available to enhance the resolution by lengthening the column if necessary (which would extend the analysis time). To obtain a sufficient number of plates, the stationary phase or support should have the smallest particle size possible. If several mobile phases are suitable, the one with the lowest viscosity should always be chosen. For a given pressure, the plate number and linear eluent velocity can be optimized by decreasing the mobile phase viscosity.

If high-speed operation is desired, the shortest possible column should be employed and perhaps even baseline resolution might be dispensed with. Optimization here proceeds at the expense of resolution. If need be, PLB can be used. The optimum speed of analysis is obtained at k' values of about 2. In HPLC it is possible to carry out analyses in seconds. However, where this may be appropriate remains an open question.

Up to now, the load capacity has not played a crucial role in HPLC. Optimizations were primarily carried out with respect to speed of analysis and resolution. Columns containing PLB were frequently overloaded with sample in order to be able to detect the components. The loss in efficiency was compensated by lengthening the column.

To achieve adequate load capacity for *preparative applications* , the column cross-section must be enlarged to accomodate sufficient stationary phase. The mobile phase (and the stationary phase dissolved in it in the case of partition systems) should be volatile to permit convenient isolation of the separated components. In preparative chromatography the resolution is always lower than in analytical applications. The analysis time is always longer than for comparable analytical separations because long columns are unavoidable.

In the "magic triangle" of chromatography

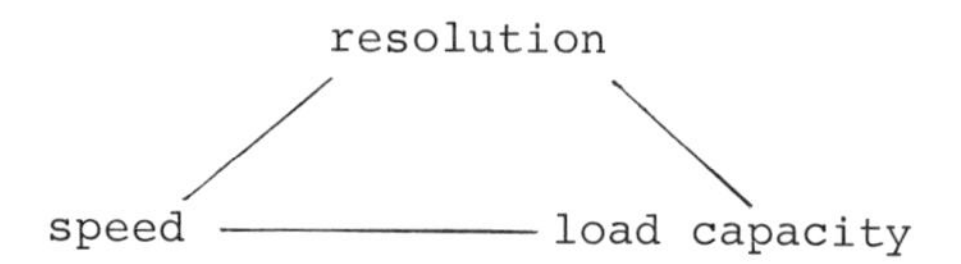

optimization can be accomplished only along any one straight line at a time, and the further is the optimization driven in one direction, the further one is removed from the two other parameters. A system that combines high resolution with high load capacity and rapid analysis has not yet been described.

References Chapter II

1. Anderson, J.R.: J. Am. Chem. Soc. *78*, 5692 (1956)
2. Kelker, H.: Ber. Bunsenges. *77*, 187 (1973)
3. Carman, P.C.: Flow of Gases through Porous Media. London: Butterworth 1956
4. Bohemen, J., Purnell, J.H.: J. Chem. Soc., London 1961, 360
5. Deninger, G.: Ber. Bunsenges. *77*, 145 (1973)
6. Halâsz, I.: Ber. Bunsenges. *77*, 140 (1973)
7. Endele, R., Halâsz, I., Unger, K.: J. Chromatogr. *99*, 377 (1974)
8. Halâsz, I., Naefe, M.: Anal. Chem. *44*, 76 (1972)
9. Giddings, J.C.: Dynamics of Chromatography. New York: Marcel Dekker 1965
10. Littlewood, A.B.: Gas Chromatography, 2nd ed. New York: Academic Press 1970
11. Halâsz, I. in: Kirkland, J.J. (Ed.): Modern Practice of Liquid Chromatography. New York: Wiley Interscience 1971
12. Desty, D.H., Goldup, A., Swanton, W.T., in: Brenner, N., Callen, J.E., Weiss, M.D. (Eds.): Gas Chromatography, New York: Academic Press 1962
13. Purnell, J.H.: J. Chem. Soc., London 1960, 1268
14. van Deemter, J.J., Zuiderweg, F.J., Klinkenberg, A.: Chem. Engng. Sci. *5*, 271 (1956)
15. Halâsz, I., Horvath, C.: Anal. Chem. *36*, 1179 (1964)
16. Endele, R.: Dissertation, Saarbrücken 1974
17. Snyder, L.R.: J. Chromatogr. Sci. *7*, 352 (1969)
18. Waters, J.L., Little, J.N., Horgan, D.F.: J. Chromatogr. Sci. *7*, 293 (1969)

19. Halász, I., Kroneisen, A., Gerlach, H.O., Walkling, P.: Z. anal. Chem. *234*, 81 (1968)

20. Halász, I., Kroneisen, A., Gerlach, H.O., Walkling, P.: Z. anal. Chem. *234*, 97 (1968)

21. Halász, I., Walkling, P.: Ber. Bunsenges. *74*, 66 (1970)

22. Deininger, G., Halász, I.: J. Chromatogr. Sci. *9*, 83 (1971)

23. Halász, I., Heine, E., in: Purnell, J.H. (Ed.): Progress in Gas Chromatography. New York: Interscience 1968

24. Halász, I., Aßhauer, J., Endele, R.: J. Chromatogr. *112*, 37 (1975)

25. Kennedy, G.J., Knox, J.H.: J. Chromatogr. Sci. *10*, 549 (1972)

26. Knox, J.H., Vasvari, J.: J. Chromatogr. *83*, 181 (1973)

27. Knox, J.H., Pryde, A.: J. Chromatogr. *112*, 171 (1975)

28. Knox, J.H., Jurand, J., Luird, G.R.: Proc. Soc. Anal. Chem. *1974*, 310

29. Knox, J.H., Saleem, M.: J. Chromatogr. Sci. *7*, 614 (1969)

30. Majors, R.: J. Chromatogr. Sci. *11*, 88 (1973)

31. Kirkland, J.J.: J. Chromatogr. Sci. *10*, 129 (1972)

32. Halász, I.: Z. Anal. Chem. *277*, 257 (1975)

33. Golay, M.J.E., in: Desty, D.H., (Ed.): Gas Chromatography 1958. London: Butterworth 1958. S. 36 ff.

34. Halász, I., Schmidt, H., Vogtel, P.: J. Chromatogr. *126*, 19 (1976)

Chapter III

Equipment for HPCL

The design of the HPLC equipment does not depend on the separation principle employed. Commercial instruments are usually designed for analytical operation, but permit preparative application on a micro scale as well. The primary impetus has been to develop modules that can be readily assembled to produce an instrument tailored to the individual's needs. Consequently, the separate components will be discussed, including the minimum standards they must meet to ensure satisfactory operation.

Today's rapidly evolving microprocessor technology has opened new dimensions in instrument capabilities in terms of function and operation (e.g., automatic sample introduction, electronic smoothing of flow pulsations, programmable detector wavelength, data processing, etc.). The current trend in LC instrumentation is proceeding in two directions. On the one hand, fully integrated compact systems incorporating all features essential (or not) for HPLC are *presently* being developed. The alternate approach would be the assembly of an instrument from individual modules that perform independently and can - if necessary - communicate with each other. At present it is impossible to predict which of these approaches will predominate in the future, although the latter could be better tailored to the needs and the funds available.

A block diagram of a typical HPLC instrument components is presented in Fig.III.1. From the solvent reservoir the pump delivers a constant flow which, depending on the type of pump, may have to be smoothed out by means of a damping device. At the point where the highest pressure occurs - usually directly after the pump - a safety valve should be inserted. From the damping unit the mobile phase flows via the sample injector to the column.

The inlet pressure of the column is measured with a manometer. After leaving the column, the sample components are sensed by a detector and monitored by a potentiometric recorder. Other monitoring devices such as integrators, etc. may also be connected to the de-

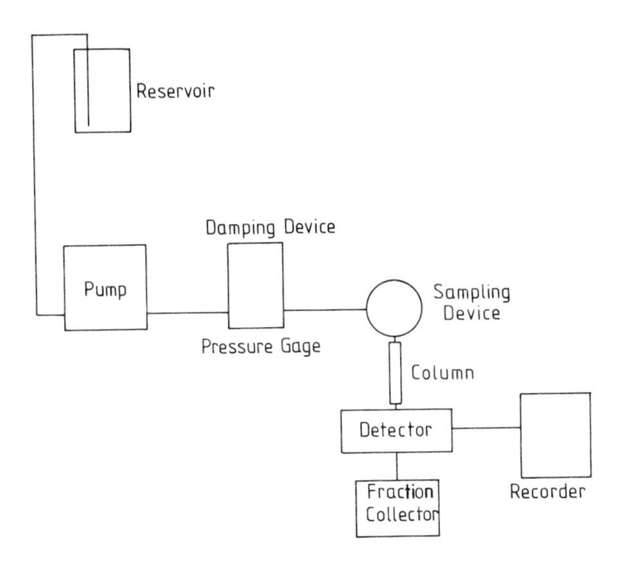

Fig.III.1. Schematic of HPLC equipment

tector. If the sample components are to be isolated, this can best be accomplished by means of a fraction collector, preferably one controlled by the recorder. The sample injector, column, and detector should be thermostatable in order to permit operation at various temperatures. The mobile phase must be adjusted to column temperature *prior* to sample introduction.

A. Solvent Reservoir – Degassing of the Eluent

For analytical applications the reservoir should have a capacity of about 1000 ml, in order to avoid frequent refilling. In some commercial instruments the reservoir is equipped with a costly device for degassing the eluent supply. In these cases the reservoir is provided with a heater, temperature regulator, and magnetic stirrer. In addition, there is a condenser for the eluent vapors to which a vacuum system is sometimes attached to accelerate the degassing. Because many eluents form explosive mixtures with air, the system must be purged with nitrogen or other inert gas, which again requires regulators and flow-measuring devices. For water-methanol mixtures, however,

degassing is essential. Whether it is preferable to do this in the reservoir itself or to degas the mixture before pouring it into the reservoir is uncertain. Indeed, the general need for degassing has been much discussed.

It is agreed that air bubbles formed on depressurization can seriously interfere with sample detection. A small back pressure, (1 - 2 atm) on the detector cell prevents the appearance of air bubbles. This back pressure can be generated simply by lengthening the detector outlet capillary.

One of the reasons for the appearance of air bubbles is poorly sealed fittings that allow air to be sucked into the system (similar to an aspirator pump). This problem can be eliminated by sealing the fittings with Teflon tape.

Decomposition of readily oxidized stationary phases or samples is certainly a rare phenomenon because of the low temperatures, but is entirely possible. Peroxide formation in the eluent can also occur. When operating with ternary mixtures, the reservoir and column should be maintained at the same temperature to avoid solvent demixing. The reservoir should be mounted or constructed in a way that permits the eluent to be changed rapidly. For some pumps it is necessary to have a certain hydrostatic pressure on the inlet valves which is easily accomplished by mounting the reservoir 50 cm above the pump.

B. Pumps

Eluent flow through the column at high pressures should be continuous and pulse-free. The pumps for analytical applications (column diameters up to 5 mm) should be able to deliver up to 20 ml/min of eluent at pressures up to 300 to 400 atm. For preparative work a higher output is required.

The following pumps are suitable for solvent delivery:

1. *Single-stroke piston pumps* with constant eluent flow
2. *Reciprocating piston pumps* and *diaphragm pumps* with a pulsating flow and *constant stroke frequency*
3. *Reciprocating piston pumps* with *variable stroke frequency*
4. *Gas-driven displacement pumps*

Systems 1,3, and 4 provide a nearly pulse-free eluent flow, whereas the pumps of system 2 require a damping device before the sample injector for smoothing the flow. The pumps are designed for either constant pressure or constant flow operation. As long as the resistance (= column) remains constant, a constant pressure will produce a constant flow (and vice versa).

1. Syringe-Type Pumps

In these pumps the piston is driven at a slow, constant rate and the mobile phase is delivered continuously. When the piston reaches its end position, the output is interrupted, and the piston is refilled by a suction stroke. The delivery time depends on the cylinder volume (between about 100 and 500 ml) and the amount delivered. The advantages of this type of pump are the absence of valves and the delivery of a constant, pulse-free flow of solvent. A disadvantage lies in the necessity of more frequent interruptions to refill at the higher flow rates. Pumps of this type are relatively expensive.

2. Reciprocating Piston Pumps and Diaphragm Pumps

These pumps deliver a continuous but pulsating flow. The sealing problems with small diameter pistons are relatively simple. Diaphragm pumps are recommended for chromatographic purposes because the parts that are contacted by the eluent can be readily made from inert materials such as stainless steel. In such pumps the piston movement is transmitted by means of a hydraulic fluid to the diaphragm and thence to the eluent. Because the piston seals contact only the hydraulic fluid (a specified high viscosity oil), the sealing problems are reduced and the reliability increased. The eluent flow is regulated by ball-check valves.

The disadvantage of this type of pump lies in the dependence of the delivered amount on back pressure caused by the dead volume and the valves. Their output decreases with increasing pressure. Therefore, in evaluating a pump it should be ascertained whether the output at maximum pressure is adequate. Diaphragm pumps are very durable and relatively inexpensive. The damping problems become relatively small when several diaphragm pumps are offset and used in tandem for the same system. By utilizing three heads, phase-shifted by 120° from each other, the pulsations in the resulting eluent stream are largely smoothed out. The pulsations of piston pumps with high piston speeds can be smoothed out more simply than those with a relatively slow piston movement.

3. Pumps with Variable Stroke Frequency

The previously described diaphragm pumps operate at constant stroke frequency. The output volume is varied by changing the stroke length. Decreasing the stroke length of a given pump chamber reduces its efficiency because it is necessary to compress large volume relative to that actually delivered. Therefore, almost all pumps developed for HPLC operate at constant piston displacement. The output volume is adjusted by means of the stroke frequency. Often the two piston movements are not sinusoidal and not exactly opposed.

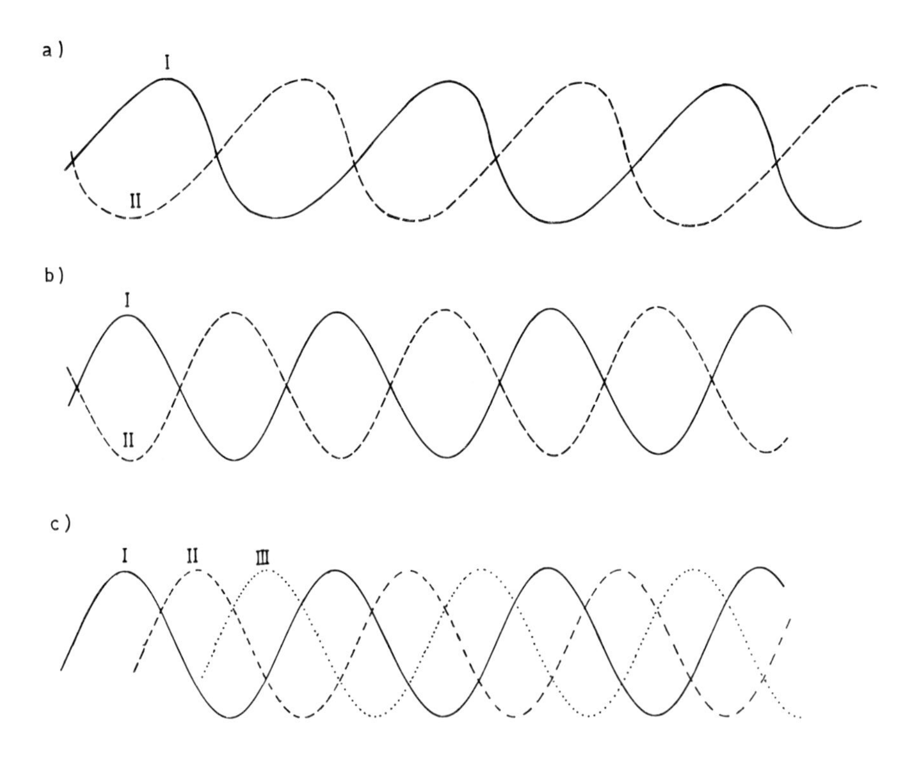

Fig.III.2. Piston movements of pumps. a) Pump with a variable stroke frequency. b) Diaphragm pump with 2 heads phase-shifted by 180°. c) Diaphragm pump with 3 heads phase-shifted by 120°

The pumps are designed so that the second piston begins to deliver before the first one has reached the end of its stroke. At this point the second piston is producing its maximum output whereas the first one is reversing and refilling. Figure III.2 compares the delivery of this type of pump with that of conventional piston pumps (two or three pistons phase-shifted by 180° or 120°).

The pulsations are barely noticeable even at low stroke frequencies. Moreover, pulsations and variations in the output can be readily smoothed or compensated electronically by measuring the pressure drop across a fixed resistance. These data are stored and a pump motor is actuated whenever deviations occur from a nominal value, so that constant pressure is maintained across the resistance and the pulsations are smoothed.

Pumps with an electronically regulated stroke frequency (driven by stepping or DC motors) are particularly well suited for installation in gradient systems.

4. Gas-driven Displacement Pumps

Gas-driven pumps, in their simplest form, consist of a plastic vessel within a pressurized chamber that is connected to a gas cylinder. In an even simpler method pressure from a gas cylinder is used to drive the eluent contained in a long tube. These pumps were used at the beginning of the development of HPLC. The maximum obtainable pressure is restricted by that of the cylinder (150 - 200 atm). Unfortunately, the gas may diffuse through the separating membrane and dissolve in the eluent at the pressures used. Therefore gas bubbles are likely to appear at the end of the column and interfere with the detector response.

Such problems are not encountered with gas-driven piston pumps, which operate by exerting a relatively low pressure (less than 10 atm) on a piston with a relatively large cross-sectional area, which in turn drives a smaller piston that delivers the carrier liquid. In this way the gas pressure can be amplified 50-fold. Up to 70 ml of eluent can be delivered with a single stroke. The solvent chamber is rapidly refilled (within a few seconds) on depressurization of the gas cylinder, resulting in scarcely any interruption of solvent delivery. The advantages of this type of pump include the ability to generate high pressures in a relatively simple manner and to produce a nearly pulse-free output. Gas-driven pumps furnish a constant eluent flow only if the pressure drop of the system does not change. The flow rate must be checked frequently because the pressure drop can be altered readily by abraded particles from the injection system. Pressure amplifiers have also been employed in HPLC. Instead of a gas, these amplifiers are driven by a hydraulic liquid delivered by a low-pressure pump.

C. Damping of the Pulsations

The eluent pulsations generated by short-stroke piston pumps must be damped before reaching the column because they may affect the detector response and deteriorate the column packing. Pulsations can be smoothed analogous to the rectification of an alternating current [1]. A damping device in its simplest form consists of capacitors and resistors connected in series. Capillaries and restrictors can be used as hydraulic resistors, and gas-filled cavities or volume-elastic containers (e.g., Bourdon tubes) as pneumatic capacitors. It is essential that such hydraulic components be contained entirely within the eluent stream or can be readily flushed out to facilitate solvent changeover. Analogous to a.c., where full wave rectification produces better results than half-wave rectification, it is easier to smooth the pulsations with two pump heads operating 180° out of phase.

Variations in the carrier velocity should not exceed 1%, and preferably 0.5%, over the entire range. This type of pulsation damping is achieved at the expense of the operating pressure. Depending on the length and width of the restriction capillaries used, up to about 40 atm of pressure may be lost as a result of the damping. Small teflon capillaries enclosed in a hydraulic fluid can also serve as pulsations damping. Pumps developed recently exclusively for HPLC incorporate an electronically controlled piston movement that largely eliminates the usual pulsations.

D. Sample Introduction

The sample is introduced either by syringe injection through the septum of an injection port into the eluent stream (as is usual in GC) or by a sample loop from which it is swept into the system by the eluent.

The sample should reach the column without any appreciable mixing with the eluent, i.e., it should be injected directly on top of the column (*on-column injection*). Furthermore, the pressure and flow equilibria should not be disturbed during sample introduction. In practice it is also important to be able to vary the sample volume readily.

1. Sample Loops

In exclusion chromatography samples are usually introduced by means of loops because the highly viscous polymer solutions are difficult to handle via syringes. The volume of such loops (between about 0.05 and 2 ml) is too large for high-speed LC. For smaller volumes the smearing of the sample in the eluent leads to an unavoidable stream profile that results in increased band broadening. Recently, sample loops with smaller volumes (1 - 20 μl) have become commercially available. Their leakage problems, which were so prevalent especially after frequent use, appear to have been solved. These new types are even suitable for pressures exceeding 100 atm. Care must be taken that these small-volume sample loops have no appreciable dead volume between the valve outlet and the column.

In addition to exclusion chromatography, sample loops are also well suited for preparative work. Also, loop sampling readily lends itself to automation, particularly for repetetive analyses of similar samples.

2. Injection Devices

The optimal sample introduction technique appears to be syringe injection through a septum into the eluent stream just on top of the column. Proper design of the injection port results in a virtually negligible increase in the dead volume. Injection ports are commercially available, and their construction has been frequently described [1,2].

Syringes with well-fitting plungers, as commonly used in GC, are suitable for operation at pressures of 50 - 100 atm, although those that can be employed up to *ca.* 600 atm are also commercially available. In addition to a tight plunger seal (which is essential for reproducible injection even at low pressures), it is important to provide a mechanical protective device that prevents plunger ejection on penetration of the membrane.

The septum itself presents the greatest problem in this type of sampling. Not only must it be able to withstand high pressures, but it must also be resistant to the common solvents. The pressure strength of septa can be enhanced appreciably by an appropriate design of the injection port [1]. It is more difficult, however, to find materials that are plastic, penetrable, self-sealing, and stable to organic solvents, yet contain no plasticizers, antioxidants, etc. that can be leached out by eluents. The latter problem, which manifests itself as the periodic appearance of certain bands or a baseline drift, may be minimized by boiling the septa in the eluent prior to use. Some septum materials become brittle from contact with organic solvents. With each injection small pieces of the septum are broken off and retained by the column. These pieces change the column permeability, as is evidenced by an increase in pressure or a decrease in flow rate. The danger of sweeping plasticizers, oligomers, etc. into the column also increases considerably from a deposit of these septum particles. Therefore, during every change of septum, the residue should be removed from the column inlet. It is advisable to place a small plug of glass wool or filter paper before the column to act as a filter, which can then be easily removed along with the septum debris. This also prevents the particles of the column packing from clogging the syringe needle during on-column injection.

Buna-N and Neoprene® are suitable septum materials for use with water, alcohols, and n-hexane as eluents. More resistant than these is Viton A® which can be readily used with aromatic eluents. However, its lifetime is very limited with methylene chloride or chloroform. In such cases septa of white silicone rubber have proved superior.

In commercial instruments some elegant solutions to sampling problems have been devised. One method involves a syringe having a long needle with a side outlet [3]. The needle is guided by means of two Teflon seals and is passed clear through the eluent stream so that it can be filled through the outlet opening outside of the injection system. To inject a sample, the syringe is retracted so that the opening is in the eluent stream directly at the top of the column. After injection the opening is removed immediately to avoid flushing the sample from the needle.

Another device consists of a syringe that remains in the eluent stream on top of the column until it is to be loaded with sample. During loading the eluent is diverted through a by-pass, but on completion, flow through the syringe is resumed, sweeping the sample onto the column.

In another system the sample is placed under ordinary pressure in a holding loop directly at the head of the column. The sample is introduced by switching valves to divert the eluent stream through the holding loop and thereby flush the sample onto the column. The additional band spreading caused by this type of sampling device is negligible even for columns packed with 10 μm particles. Automatic sampling devices are also available now for various instruments.

In some cases *stop-flow injection* remains as the only possibility. In this technique the eluent flow is interrupted and the sample is injected under ordinary pressure. The attendent additional zone broadening is negligible. After sample application the eluent flow is resumed, pressure is built up, and the separation is initiated. A certain amount of time elapses, however, before the pressure profile within the column is re-established. Hence, stop-flow injection does not meet the stipulation that sampling should not disturb the pressure and flow equilibria of a system, and the sample retention times obtained using this technique are incorrect. This drawback can be circumvented, however, by the addition of an internal standard and the determination of relative retentions; this is essential for quantitative work. Another disadvantage of this technique is the possible change in the packing density of the column caused by the pressure surges.

The volume of a sample will depend on its solubility in a particular eluent. However, it should not be too large because this can lead to additional band broadening. Danger of this is particularly acute when the sample volume approaches the same order of magnitude as the mobile phase volume in the column. For analytical work the sample volume should be kept as small as possible, i.e., the sample should be injected as concentrated as possible. In contrast, for preparative operation the head of the column may be overloaded with a concentrated sample solution.

E. The Column

1. Column Materials

Columns are generally made of stainless steel tubing. Glass and tantalum tubings are also used. Glass tubing possesses the great advantage of allowing the column packing to be observed constantly. However, it generally cannot be used above 70 atm pressure. Jacketing the glass tubing in a steel tube and pressurizing the annular space enables operation at higher pressures [4]. Glass tubing has a smooth inner surface. In this respect, columns made of tantalum are supposed to be equivalent to those of glass [3]. Tantalum tubing is very hard; the connections must be sealed on.

Depending on the manufacturing process, the inner surface of stainless steel tubing is grooved and more or less rough. This roughness can be removed by drilling out or polishing the inner tube surface. Such columns can be packed more reproducibly, irrespective of the original manufacturing process used (hot- or cold- or precision-drawn, etc.). The separation efficiency obtained with such smoothed tubing may be up to a factor of 10 better than that of untreated tubing; the packing reproducibility rises from 30% to almost 90% [5].

The properties of the most important tubing materials may be summarized as follows:

Glass:	smooth inner wall; transparent; inert; useable up to about 70 atm.
Stainless Steel:	relatively corrosion resistent; can be passivated; no pressure limitations.
Tantalum:	smooth inner wall; largely inert; hard; difficult to shape.
Copper:	easy to work; subject to corrosion.

Therefore, glass tubing is to be preferred, especially for aqueous salt solutions such as acetate or citrate buffers, but it turns out be more convenient to use stainless steel. The internal diameter of analytical columns is generally 2 - 4 mm. For > 30 μm particles 2 mm i.d. columns are used almost exclusively. With ~10 μm particles 3 - 4 mm i.d. columns appear to pack more reproducibly and yield better efficiency than those with smaller i.d. [6].

2. Column Shape

The shape of a column is determined largely by the size and shape of the available thermostat. Straight columns are easier to pack than coiled ones. Due to the slow radial diffusion in the liquid mobile phase, equalization of the stream profiles in coiled columns occurs much more slowly than in gas chromatography. Packed columns should be bent to the largest radius possible. Even then some particles may be crushed to the detriment of the column permeability. For $d_p \leqslant 10$ µm straight columns are used exclusively, and the lengths are commonly 10 - 50 cm.

3. Column Connections

A metal frit or glass or metal wool are to be preferred over a porous Teflon® washer for plugging the ends of a column. If necessary, filter paper ("intended for the filtration of fine precipitates") can be

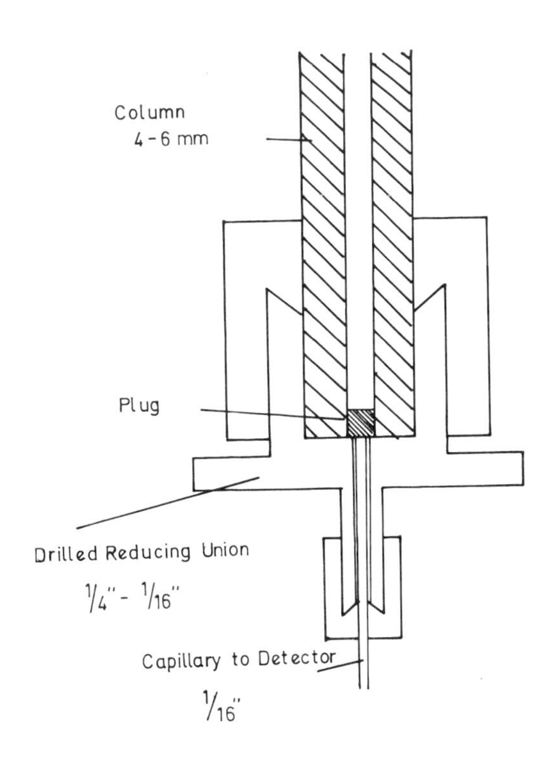

Fig.III.3. Column end-fitting (schematic)

inserted at the column outlet. On the one hand, the flow resistance of the column end fitting should be minimal, on the other, the unavoidable fines (< 1 μm) of a column packing should be prevented from reaching the detector or even clogging the narrow connecting capillaries. As has been emphasized repeatedly, all unnecessary dead volume in the connections from the column to the injection block and detector, as well as between separate column sections, should be avoided [1]. All passages and connections should be drilled out to achieve seamless transitions between the injection block and the column, the separate column sections, and the column and detector.

Fig.III.3 shows the connection of a 6 mm column to a 1/16" capillary to the detector. A commercial reducing union was drilled out to permit the detector connection to be attached directly to the end of the column.

4. Column Packing

The method employed for column packing depends on the particle size of the stationary phase. Two fundamentally different methods are used:

Particles with diameters > 20 μm can be dry-packed. In this method the column is clamped in a vertical position and the packing is added in small increments while the tube is tapped and vibrated. The portions are measured so that the packing height increases only a few millimeters after each addition [7,8]. If the column is also tamped on a firm support during this process, efficient and reproducibly packed columns are obtained.

This method is also suitable for packing supports coated with a liquid phase. If the particles are somewhat sticky, better results are frequently obtained by increasing the packing density by tamping with a tightly-fitting rod. With any packing method it is essential to avoid crushing the particles and producing fines by mechanical abrasion.

The following are guideline values: A well-packed column containing totally porous particles (d_p = 30 - 40 μm) should exhibit a plate height of around 1 mm at a linear velocity of 2 cm/sec. Under the same conditions PLB show lower H values (0.2 - 0.5 mm).

For ~ 10 or 5 μm particles dry-packing is unsuitable. They are best packed by a slurry technique into a column containing the eluent. The suspension should remain stable throughout the packing procedure, i.e., the particles should not sediment or agglomerate. Stable suspensions are obtained if either

a) there is no difference in the densities between the particles and the dispersing medium; the *balanced density slurry method* is based on this, or

b) the viscosity of the dispersing medium is so high that the sedimentation rate of the particles is minimal; the *high viscosity method* is based on this.

The balanced density method has been described in detail by various authors [5,9-12] and suitable packing vessels (suspension reservoirs) have been reported [9,12]. Only bromine- or iodine-containing hydrocarbons can be used to suspend silica gel, whose actual density is 2.2. Tetrabromoethane is used almost exclusively, carbon tetrachloride being added to reduce the density to the required value. It is also desirable to add about 10% of a polar compound (dioxane, methanol) to prevent aggregation of the particles.

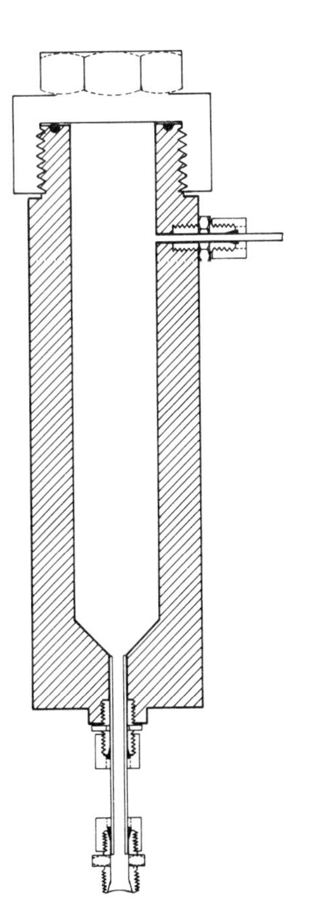

Fig. III.4. Sketch of a packing vessel. Volume: *ca.* 100 ml for analytical columns (30 cm, 4 mm i.d.), ca. 200 ml for (30 cm) columns up to 10 mm i.d. The outlet tube of the packing vessel should have the same cross-section as the column to be packed; the fitting of the outlet tube should be mated with that of the sampling device

Fig.III.4 shows one version of a packing vessel. The total volume should be about 100 - 150 ml to provide enough space for the packing of micro-preparative columns. To pack a 30 cm long 4 mm i.d. column requires about 2 g of silica gel in 50 ml of the dispersing medium (e.g., 20:15:15 tetrabromoethane-dioxane-carbon tetrachloride). This suspension is carefully poured into the packing vessel (cf. Fig. III.4). The column along with a *ca.* 2 cm extension of the same diameter is connected to the packing vessel and filled with the eluent. The extension is also packed but is not used in the separations. The suspension is covered with an eluent (e.g., heptane) that is less dense than, and miscible with, tetrabromoethane. After closing off the packing vessel, it is important that the pump be connected in such a way that no air bubbles enter the system. The suspension is then pumped into the column. The pressure applied during packing should exceed the highest pressure anticipated during actual operation. In this case the conclusion of the packing process (displacement of the tetrabromoethane by heptane) is indicated by a drop in pressure (due to the lower viscosity of the heptane). The column should be flushed a few minutes longer to remove the dispersing medium completely and to compress the packing. The use of an organic solvent as a displacing agent instead of water (which is immiscible with tetrabromoethane) has the advantage that the column can be used immediately without any further conditioning (e.g., the tedious removal of water).

Tetrabromoethane readily splits off bromine or hydrogen bromide which are not completely removed by flushing with heptane, water, methanol, etc. Chemically bonded phases may be decomposed or modified by packing from tetrabromoethane. In such cases it is therefore preferable to use the *viscosity method*, which utilizes a high-viscosity liquid (such as paraffin oil, cyclohexanol, etc.) as the suspending medium, but otherwise identical apparatus and procedure. A mixture of cyclohexanol and isopropanol (1:2) is a good suspending liquid for chemically bonded phases as well as for silica. A combination of the two methods (density and viscosity) can also be employed. The more rapidly the suspension is handled (the less time there is for particle settling), the less important is the suspending medium. Hence, if the slurry handling steps are carried out expeditiously, pure isopropanol ot tetrachloromethane may be used as the suspending medium. Virtually all stationary phases can be packed by one of these two suspension techniques if the density and/or the viscosity of the dispersing medium is adjusted so that the suspension remains stable at least until the column has been packed.

A suitable mixture for packing alumina consists of 90:10 v/v tetrabromoethane-dioxane. Pre-swollen stationary phases for exclusion chromatography, ion exchangers based on polystyrene, etc. can only be packed from the medium in which they were swollen and in which the separation is to be carried out. A change in eluent may alter the degree of swelling and hence affect the column packing. Supports coated with a liquid phase intended for partition chromatography *cannot* be packed in this manner because the liquid phase may be stripped off during packing or conditioning. Only pure supports can be packed by this method; coating with liquid phase can only be carried out subsequently *in situ* (cf. Chapter VII).

3. Characterization and Testing of Columns

As has been pointed out in Section II.F, the attainable plate height of a column is a function of the particle size of the stationary phase. The smaller the particle size, the smaller is the plate height or the greater is the plate number for a given column length. However, the assumption in this case is that all columns can be packed equally well. As particle size is decreased, especially < 10 µm, greater demands are placed on the equipment. The contribution from extra-column band spreading may be considerable, even to the extent that the H values show no dependence on particle size. The extra-column band spreading generated in the sampling device, detector, and connections can only be determined with great difficulty (a high-speed recorder, for example, is essential). Theoretical considerations, such as the real band broadening within the column, are of little value to the practical chromatographer. Instead, he requires a certain number of plates to carry out his separation and is only interested in establishing whether or not he will achieve his goal with his column and all attendant instrumentation.

The efficiency of a column and the minimum band broadening that is attainable can be evaluated from chromatographic data by the following procedure [14].

a) *Asymmetric peaks* , especially the unretained peak and those weakly retarded ($k' < 3$), are typical for a poorly packed column if instrumental effects can be excluded. If only the unretained peak and those with $k' < 1$ are asymmetric, the problem may also be instrumental in nature. This can only be resolved by means of an H *vs* u plot.

If there are deviations from a straight line at linear velocities < 3 mm/sec, the problem may be attributed to instrumental difficulties with a high degree of probability.

To determine peak asymmetry, a perpendicular is dropped from the peak maximum to the baseline w (cf. Fig.II.1) and the ratio of the larger to the smaller baseline segments between the tangents drawn through the points of inflection is calculated. If the asymmetry factor thus calculated exceeds 1.5 (or 2.0 if the square of the asymmetry factor is used [5]), the column should be discarded inasmuch as "tailing" or "leading" diminishes the resolution and in micro-preparative separations contaminates the resolved components.

b) *The retention time* of the unretained peak t_o, from which the linear velocity can be computed as well, is also necessary for the characterization of a column. In HPLC it is often difficult to decide whether a substance is retarded or not, particularly if only a UV detector is available. With aliphatic hydrocarbons, such as heptane, as the eluent, tetrachloroethylene or even carbon tetrachloride may be assumed to be inert. If a differential refractometer is available, the dead time can be determined reliably with a lower homolog of the eluent. Benzene is not retarded in methylene chloride as eluent; neither, of course, is n-heptane.

For chemically bonded stationary phases with water or water-organic solvent mixtures the accurate determination of the dead time becomes more complex. In aqueous mixtures either water or the organic solvent may be regarded as inert (differential refractometer). Deuterium oxide is also not retarded in water. However, problems arise when using salts or dissociated substances, as these may even elute before the unretained peak as a result of exclusion effects (i.e., the Donnan potential). In such cases t_o can only be determined with nonpolar eluents and unretained compounds before the column is converted into the polar system. It can be assumed that the porosity of a column is not altered by eluent changes.

As a check on the measurements, t_o can be estimated by rearranging Eq.(11)

$$t_o = \frac{L\varepsilon_T r^2 \pi}{F}$$

where L is the column length (cm), r the column radius (cm), and F the flow rate (cm^3/sec). The total porosity ε_T of totally porous supports (regardless of the packing method) is 0.84 $\pm$ 5%. For chemically bonded phases, including ion exchangers, ε_T is always smaller, and 0.75 is a good approximation. In the limiting case $\varepsilon_T = 0.42$ (e.g., for solid glass beads).

c) The *particle size* of a stationary phase quoted by the supplier does not necessarily correspond with the effective "hydrodynamic" particle size that affects the efficiency and permeability. As explained in Chapter II.F, this hydrodynamic particle size can be determined via the column permeability or the pressure drop required to achieve a certain flow rate. By rearranging Eq.(29) one obtains [13]

$$d_p = 41 \sqrt{\frac{F\eta L}{r^2 \Delta p}} \ .$$

d_p is obtained in μm if F is expressed in cm^3/min, L in cm, η in cP, r in mm, and Δp in atm. (The conversion factors for the cgs system are contained in the constant).

d) The *plate height* H to be expected for well-packed columns can be estimated from the following approximation [14], using the particle size as defined above and the linear velocity u computed from the dead time:

$$H = 3\ dp + \frac{6}{u} + \frac{d_p^2}{16} \cdot u \ .$$

This equation holds only for nonpolar eluents of low viscosity, such as n-heptane and methylene chloride, and low molecular weight samples, e.g., benzene and substituted derivatives such as nitrobenzene, nitroaniline, nitrophenol, etc. In contrast to the cgs system, this equation gives H values in μm when d_p and u are substituted in μm and mm/sec, respectively. This equation permits only the H value to be calculated, but does not allow one to differentiate between, or to calculate, the individual A, B, and C terms.

A packed column yielding symmetrical peaks is considered satisfactory if its H values are up to 1.5 times the value calculated above. Based on experience, for nonpolar stationary phases (reversed phases) and polar eluents (e.g., water) the H values are double those calculated by the above equation.

Band broadening rises with increasing flow velocities if those less than 1 - 2 mm/sec are disregarded. However, with 5 μm particles the B term becomes noticeable even at these velocities (cf. II.F).

Table III.1 summarizes the efficiencies attainable with various

Table III.1. Comparison of particle size, efficiency, and analysis time

Particle size	Type of particle	Packing method	Eluent velocity [mm/sec]	Plate height [µm]	Column length [cm] required for *ca.* 3000 plates	Pressure drop [atm] [1] required for *ca.* 3000 plates	Max. Analysis time [min] [2]
"40 µm"	porous	dry	20	1500-3000	900	450	30
			10	900-2000	600	150	40
"40 µm"	PLB	dry	10	400-700	200	50	14
"30 µm"	PLB	dry	10	200-400	120	55	8
"10 µm"	porous	suspension	10	100-150	45	180	3
			5	60-100	30	60	4
			1	45-75	25	10	17
" 5 µm"	porous	suspension	10	50-70	20	320	1.4
			5	30-50	15	120	2
			2	20-40	12	40	4
			1	15-30	10	15	7

[1] with n-heptane at the given eluent velocity

[2] the k'value of the last peak = 3

particle sizes and packing methods. To illustrate the characteristics of different particle sizes, the data are based on a column length required to produce 3000 plates (which is satisfactory for many routine separations). The required pressure drops are also based on these column lengths. The analysis time (last column) always decreases with diminishing column length, but whether it is always practical to work at a high velocity will be made clear with the following example: If one works with 5 μm particles at 2 mm/sec, a pressure of only 40 atm and a 12 cm column are required. The analysis takes only about 4 min, compared to 1.4 min for a column almost twice as long with an approximately eight-fold greater pressure drop.

e) The *load capacity* of column is usually about 10^{-4} g sample/g stationary phase. If this sample size is exceeded, the plate height and retention time (k' value) become dependent on the sample size, and qualitative analysis, i.e., identification via retention times, is then no longer possible. An experimental curve of the determination of the load capacity is presented in Fig.VI,2.

F. Thermostating

Frequently, it is necessary to monitor column temperature and to maintain it constant. Air thermostats with rapid air circulation are preferred because of their simple construction. However, an additional effective heat exchanger must be built into the instrument ahead of the column in order to bring the mobile phase to the thermostat temperature. Because of their rapid response and adjustment times, air thermostats have definite advantages over the slowly-responding liquid thermostats which, however, have lesser heat transfer problems. The danger of an explosion in using air thermostats with an open heating element is mentioned only in passing. A device for purging the thermostat compartment with nitrogen should be available for safety's sake.

G. Measurement of the Flow Rate

The flow rate should be checked as often as possible. It is simplest to collect a measured volume and to note the elapsed time. The principle of the siphon can also be used [15]. In this case the eluent is allowed to flow into a siphon having a definite volume. On emptying the siphon an electrical pulse is produced by intercepting the light beam on a photocell, which then marks the recorder paper. The flow rate can be calculated from the known volume of the siphon and the time which can be deduced from the distance between the markings. Since such a device can accompany a separation continuously, constant control of the flow rate is possible.

The rotameters usually employed in GC must be standardized in LC at each temperature for every solvent because of the large differences in the viscosity and its strong temperature dependence. Consequently, they are little used. These methods integrate the flow rate over a long period. A method to determine short-term changes (during the elution of a single peak) has been proposed [16].

H. Fraction Collectors

As a result of the high sensitivity of detectors, devices for collecting individual fractions merely for detecting the sample components have become virtually unnecessary in analytical HPLC. Only for certain measurements, such as radioactivity, may it become necessary to collect individual fraction after certain volume or time intervals. For preparative work, of course, a fraction collector is indispensable. As in GC, it is preferable that the fraction collector be recorder-controlled in order to reduce the number fractions to be worked up. In addition to the fraction collectors known from classical column chromatography, smaller devices for 10 - 20 fractions have been specially developed for HPLC and are available commercially.

I. Recorders

Chromatograms are recorded with potentiometric recorders that must be adapted to the particular detector. The response should be 0.5 sec or better for full scale to avoid distortion of rapidly eluting peaks. Since the elution times are highly variable, the chart speed should be readily adjustable. Integrators and calculators can be used as in GC.

J. Instrumentation for Gradient Elution

The composition of the eluent can exert a profound effect on the outcome of a separation. By changing the eluent composition continuously, markedly shorter elution times may be achieved. This is particularly important for broad-range mixtures where not only the early-eluting peaks should be well resolved, but even the strongly retarded ones should emerge as sharp zones in considerably less time. Hence, gradient elution can reduce the analysis time considerably and enhance the detection sensitivity significantly. The resolution, however, is usually diminished. This is explained in detail in Section VI.III. Only the instrumental problems associated with gradient elution will be considered here.

The majority of the modern detectors respond to changes in the eluent composition, as will be discussed in detail later. Here it is only mentioned that baseline drift cannot be eliminated even by the use of a differential signal and a reference column ostensibly identical to the analytical column, for it is virtually impossible to pack two columns with completely identical hydrodynamic resistances. The mixing of eluents poses certain problems. Most simply, the eluents can be mixed on the low pressure side, using classical LC devices [17,18]. A short-stroke pump can then be used to deliver the solvent. Long connections between the mixing chamber and column as well as the mixing occurring in the pump and damping units may contribute to poor reproducibility. An automatic mixing apparatus with 20 reservoirs for low-pressure gradients has been described [19].

The equipment for the preparation of elution gradients on the high pressure side is more elaborate. Each solvent component requires its own delivery system. The geometry of the mixing chamber is a crucial factor in ensuring thorough mixing of the solvent components prior to reaching the column. Hence the volume between the mixing chamber and column should be minimal, and there should be no branchings in the eluent path.

There are gradient accessories for most commercial instruments. Elution gradients can be most simply prepared by means of two contra-rotating syringe pumps whose outputs vary reciprocally. This is achieved either by having both pumps feeding into the same mixing chamber or one pump feeds into the piston of the other, which serves as the mixing vessel and is additionally mixed with a magnetic stirrer. Some deviations in gradient reproducibility may originate in eluent compressibility [20].

A gradient can also be prepared with a single pump. To do this, the stronger solvent component is placed in a holding coil, whereas the weaker eluent flows directly from the pump to the column via a mixing chamber. By appropriate valve timing the pump may be connected directly to the column or indirectly via the holding coil. The gradient is generated by decreasing the opening time of the valve between the pump and column and correspondingly increasing that of the valve between the holding coil and column. Thorough mixing of the two plug-shaped eluent streams thus produced is essential.

Gradients can also be generated very simply with two-headed diaphragm pumps in which each head delivers a different solvent and the piston strokes are displaced reciprocally by means of a stepping motor. The resulting eluent flow is kept constant by monitoring with two flow meters that feed back to the stepping motors. Difficulties associated with the reproducible and uniform displacement of the piston strokes and hence with the precision of the gradient composition can be overcome electronically.

Modern HPLC pumps where the flow rate is adjusted via stroke frequency provide additional possibilities. Two pumps can be controlled electronically so that their total output remains constant, but that of each one can be varied reciprocally. In this way any desired gradient programm can be produced from two components.

It is advisable to verify the proper functioning of the gradient elution equipment because the mixing of eluents of different density and viscosity is often accompanied by contraction in volume, anomalous viscosity changes, heat evolution, etc. The following tests will reveal most of these phenomena and their effects on gradient mixing [21]. By using pure heptane and heptane containing 0.1% tetrachloroethylene (which is UV-active) as eluents A and B, a linear rate of programming will disclose deficiencies in the pump and/or gradient generator, and the presence of poorly swept volumes by deviations from a linear rise in the baseline. Gradient delay between program initiation and the appearance of eluent B at the end of the column, which is caused by dead volumes in the mixing chamber, damping devices, etc., will also be revealed by a delay in the baseline rise.

The effect of differences in eluent density on instrument performance can be checked with heptane-methylene chloride gradients (these solvents have approximately the same viscosity). Density differences affect the functioning of the ball-check valves and the mixing of the gradient components in the mixing chamber.

The effect of viscosity changes on instrument performance can be tested with a heptane-isopropanol gradient. Deviations from constant flow due to variation in the back pressure and in the mixing behavior of both solvents may be observed. These tests can be performed with or without the column and are valid for polar as well as nonpolar phases.

The greatest demands on the equipment are made by water-methanol gradients which are frequently used with nonpolar stationary phases. Not only are density changes involved, but the viscosity also passes through a maximum (cf. Fig.VI.11). The heat of mixing, volume contraction, and degassing pose additional problems. The density and viscosity effects can be ascertained by the above tests. Degassing of the individual components, e.g., under reduced pressure in an ultrasonic bath prevents the appearance of gas bubbles. The heat of mixing brings about changes in the k' values and band widths. The volume contraction (max. *ca.* 4%) must be taken into account for the accurate determination of elution volumes.

Finally, there is still no detector that is completely insensitive to changes in the eluent properties during gradient elution, although this problem has been minimized in some instruments.

Programming alternatives to gradient elution include eluent velocity (via the column inlet pressure) and the separation temperature [22]. If the output is adequate, the volume flow rate of the eluent increases in accordance with the rise in the column inlet pressure. Every programmed increase in the output volume of the pump leads to a decrease in the analysis time. Most detectors (particularly RI and UV) do not respond to increases in the eluent flow through the sample cell within certain limits (0.5 - 8 ml/min). Hence, flow programming is suitable for reducing the analysis time.

Temperature programming (continuous increase in the separation temperature) is not so universally applicable. First, only the UV detector is relatively unaffected by temperature variations, the differential refractometer being entirely unsuitable under these conditions. Second, a temperature rise disturbs the equilibrium between the water adsorbed on the adsorbent surface and that dissolved in the eluent (cf. VI,E.2). As in GC, a linear temperature gradient corresponds to an exponential pressure or flow gradient.

In addition to heating the column, the eluent must also be raised to the desired temperature *before* reaching the column. A satisfactory heat exchanger consists of a *ca.* one-meter, 0.25 mm i.d. capillary placed in the same thermostat as the column. (Jacketing like that of a Liebig's condenser is sufficient for recirculation thermostats).

K. Safety Measures

The pressure generated in instruments poses no danger because no energy of compression is stored in the noncompressible liquids. However, the high flammability of most eluents should be borne in mind. A spray of eluent may emerge from any improperly tightened place, especially from the septum of an injection block. Air thermostats should be equipped with an inert gas purge system. It is highly recommended that ovens be equipped with safety switches to prevent overheating, thereby avoiding the high pressures of the supercritical region of eluents.

A safety valve serves to prevent damage to instrument components due to excessive pressures generated by clogging. It is best to install the safety valve where the highest pressure occurs - directly after the pump.

The health hazards of certain organic solvents, especially the carcinogenic properties of some, should not be overlooked.

References Chapter III

1. Halász, I., Kroneisen, A., Gerlach, H.O., Walkling, P.: Z. Anal. Chem. *234*, 81 (1968)
2. Pearce, B., Thomas, W.L.: Anal. Chem. *44*, 1107 (1972)
3. Ecker, E.: Chemiker Ztg. *95*, 511 (1971)
4. Stahl, K.W., Schuppe, E., Potthast, H.: GIT *17*, 563 (1973)
5. Aßhauer, J., Halász, I.: J. Chromatogr. Sci. *12*, 139 (1974)
6. Boehme, W.: Diplomarbeit Saarbrücken 1973
7. Karger, B.L., Conroe, K., Engelhardt, H.: J. Chromatogr. Sci. *8*, 242 (1970)
8. Halász, I., Naefe, M.: Anal. Chem. *44*, 76 (1972)
9. Kirkland, J.J.: J. Chromatogr. Sci. *10*, 593 (1972)
10. Kirkland, J.J. in: Perry, S.G. (Ed.): Gas Chromatography 1972. p. 39 ff. Barking, Essex, England: Applied Science Publ. 1973
11. Majors, R.E.: Anal. Chem. *44*, 1722 (1972)
12. Strubert, W.: Chromatographia *6*, 50 (1973)
13. Halász, I., Schmidt, H., Vogtel, P.: Chromatogr. *126*, 19 (1976)
14. Halász, I.: Z. Anal. Chem. *277*, 257 (1975)
15. Schneider, H., Rössler, G., Halász, I.: Chromatographia *6*, 237 (1973)
16. Halász, I., Vogtel, P.: J. Chromatogr. *142*, 241 (1977)
17. Snyder, L.R.: Chromatographic Rev. *7*, 1 (1965)
18. Engelhardt, H., Elgass, H.: J. Chromatogr. *112*, 415 (1975)
19. Scott, R.P.W., Kucera, P.: J. Chromatogr. Sci. *11*, 83 (1973)
20. Martin, M., Guiochon, G.: J. Chromatogr. *112*, 399 (1975)
21. Elgass, H.: Ph. D. Thesis, Saarbrücken 1978
22. Wiedemann, H.: Ph. D. Thesis, Saarbrücken 1973

Chapter IV

Detectors

The composition of the column effluent is continuously monitored by a detector. Unfortunately, there is still no universally applicable LC detector. The physicochemical properties of the mobile phase differ very little from those of the sample components, thus requiring either very *specific* detectors (such as a UV detector) or those capable of measuring extremely small differences in the bulk properties by a differential technique (such as a differential refractometer). A suitable detector must be selected for each sample-eluent combination. Therefore, at least two different detectors should be available for each instrument. One of them should be a differential detector that responds to the bulk properties of the eluent (e.g., refractometer, dielectric constant) and the other should be specific in its response to the substances being monitored. The latter type includes the UV as well as polarographic and radioactivity detectors. The more specific the detector becomes the more one is restricted in the choice of the mobile phase.

The two most frequently used detectors in LC today are the UV and differential refractometer. Ultraviolet detectors are the most sensitive for samples having relatively high absorption coefficients at the appropriate wavelength. However, the choice of eluent is then restricted by the necessity of it being completely transparent at the detector wavelength. Differential refractometers are very sensitive to temperature and pressure fluctuations. Both types of detectors respond to the sample concentration in the eluent.

The following criteria are used for the characterization and description of detectors:

The *noise level* governs the lowest detection limit. A chromatographic peak can only be recognized as such if its height is at least twice that of the highest noise peak. In addition to the noise from purely electrical sources, air bubbles and impurities in the eluent may also be causes of this phenomenon.

A *drift* in the baseline is undesirable. Its primary causes are slow changes in the ambient temperature, the flow rate, or stripping of the stationary phase from the column.

In considering the *sensitivity* , distinction must be made between the absolute and the relative sensitivity of a detector. The former is a function of the instrument design, the measuring technique employed, and the noise level; the latter depends on the amount of a certain substance that is just detectable under a definite set of chromatographic conditions. The sensitivity is one of the most important characteristics of a detector. However, there are other factors to consider, such as band spreading in the detector, dependence of the response to external parameters, and the convenience of servicing. For *quantitative analysis* , the linearity of the response plays an important role. Unfortunately, not all recorders respond with total linearity throughout the range of application.

If several detectors are utilized, it is advantageous to connect them in series in the order of their increasing dead volumes. However, the pressure stability of the cell must be taken into consideration because with some detectors (e.g., the heat exchanger of a differential refractometer) a marked pressure drop develops, especially at higher flow rates.

A. UV Detectors

Because of their relatively low susceptibility to temperature and flow rate fluctuations, UV detechtors are widely used. Most of these instruments operate at a single wavelength - 253.7 nm, the intense band of a low pressure mercury lamp. In some instruments a band at 280 nm can be excited by introducing a suitable phosphor and permitting the light emitted to serve as the source. Most instruments are constructed to operate at only one of these wavelengths but a few allow operation at various wavelengths.

In principle, photometers or spectrophotometers, which permit measurement at any wavelength, can also be used. Especially good resolution is not essential, and band widths of 10 - 20 nm are entirely adequate. Recording double-beam spectrophotometers may be employed [19], provided the beams can be restricted to an aperture of about 1 mm in diameter without increasing the noise excessively. If an instrument conforms to these requirements, the UV spectrum of a sample component can be recorded by momentarily stopping the eluent flow as the sample band passes through the cell. Such interruption causes little additional band broadening because, as mentioned previously, the diffusion is very slow. Two recorders can be used very effectively with this arrangement. One records the usual chromatogram (absorbance *vs.* time at constant wavelength), while the other plots the spectrum (absorbance *vs.* wavelength) during the interruption of the solvent flow. Rapidly-scanning spectrophotometers capable of recording the entire spectral region every 20 sec have also been employed [50,51]. The value of this approach is questionable, however, as UV spectra provide only a limited potential for sample identification and LC-UV detectors never exhibit the full resolution attainable by normal UV spectroscopy.

The UV cells should have an optical path length of 5 - 10 mm and very small volumes (5 - 8 μl). Therefore, the aperture must also be kept small (~ 1 mm) for a path length of 10 mm. Fig.IV.I. shows a typical arrangement. It is important that the entire cell be thoroughly swept by the flowing medium and thus flushed well. The cell design should minimize turbulence even at higher flow rates so that the noise level is independent of the flow. An H-shaped cell has also been described [1].

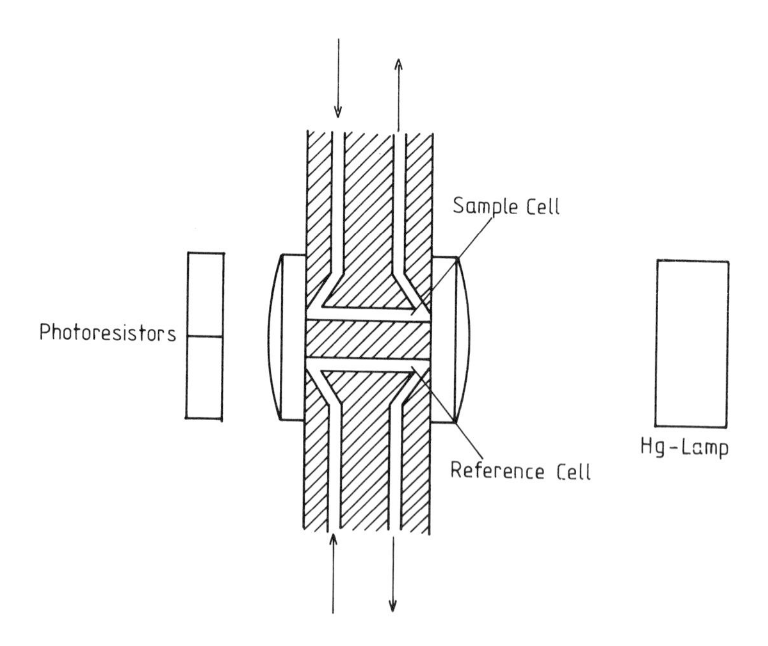

Fig.IV.1. Schematic of a UV-detector cell

The instruments may be single- or double-beam (which contain a reference cell). It is not essential, however, that the reference cell be filled or continuously flushed with eluent. On the other hand, reference cells filled with mobile phase that is free of air bubbles facilitate compensation, particularly if the eluent itself absorbs slightly at the wavelength used. The compensation for changes in the solvent composition by means of a constant reference flow having the identical composition is hardly feasible because the equalization of both flows (even with the aid of a reference column) is exceedingly difficult.

The disadvantage of UV detectors lies in their specificity. Only molecules that absorb in the UV region near the wavelength of the detector can be monitored. Due to high detector sensitivity, however, it is not essential to measure at the absorption maximum, for even the side of a band furnishes adequate sensitivity. At 254 nm all compounds containing an aromatic ring can be detected. This is also true of most ketones and aldehydes whose absorption bands extend into this region. Condensed aromatics, whose principal absorption is displaced toward longer wavelengths, exhibit sufficient absorption and can be detected with adequate sensitivity.

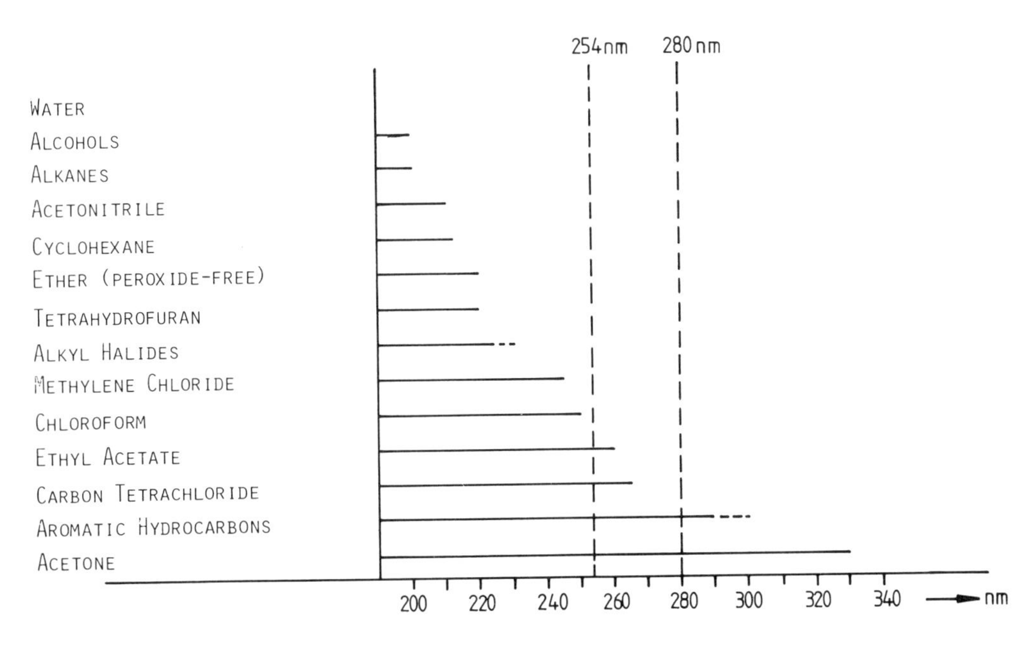

Fig.IV.2. The UV-transparency of important eluents

Catalogs of spectra or texts on UV spectroscopy provide data on the absorption maxima and molar absorptivities of substances to be measured. They should always be consulted.

The choice of a mobile phase is also restricted. Fig.IV.2 presents the regions of transparency of the important mobile phases. The data refer to purified solvents. The impurities in ordinary commercial products shift the transparency region to longer wavelengths. Hence, spectroquality solvents are recommended.

The *sensitivity* of *UV detectors* depends strongly on the molar absorption coefficients of the sample components. In the region of the wavelengths used (254 or 280 nm) these coefficients may vary between about 20 for saturated carbonyl compounds and several factors of 10 000 for aromatics, heterocyclics, etc. The noise level of almost all UV detectors lies at about 10^{-4} absorbance units (AU), although in recently developed ones it has been reduced to around 10^{-5} AU. The minimum detectable concentrations may be calculated by means of the Beer-Lambert law, assuming that the band should have an absorbance twice the noise level.

Table IV.1. contains absorption coefficients taken from the literature [2]. The last column gives the amount detectable if it were dissolved in the 8 µl volume of a detector cell (1 cm path length). However, because of the dilution that occurs during the chromatographic process, this amount of sample in a peak at the end of the column would be distributed over a far greater volume. Depending on the k' value and the column packing, this volume could be between 10 and 1000 µl,

Table IV.1. Theoretical Detection Limits of UV-Detectors

Substance	ε [2]	Detection limit (for a noise level of $2 \cdot 10^{-4}$ AU) (g/ml)		Amount of sample in an 8 µl cell[a] (g)
sat. carbonyl compound MW 100	20	$2 \cdot 10^{-6}$	2 ppm	$1.6 \cdot 10^{-8}$
Benzene	200	$1.5 \cdot 10^{-7}$	0.1 ppm	$1.2 \cdot 10^{-9}$
Benzaldehyde	11 000	$3.8 \cdot 10^{-9}$	4 ppb	$3 \cdot 10^{-11}$
Anthracene	220 000	$3.5 \cdot 10^{-10}$	0.3 ppb	$2.8 \cdot 10^{-12}$

[a] The eluent density was arbitrarily chosen as 1 g/ml

or perhaps even greater. Consequently, at least a ten to hundred fold amount of that given in Table IV.1 must be injected in order to obtain a perceptible signal.

A UV detector is very sensitive and selective. Thus, it is applicable to gradient elution only if the eluent has no UV absorption in the region of the wavelength used. However, even then baseline drift may appear because of changes in the refractive index, depending on the particular cell design. Through optical methods and suitable cell design this effect of the refractive index can be suppressed.

B. Differential Refractometer

A differential refractometer measures the bulk refractive index of a sample-eluent system. In order to obtain adequate sample response, the refractive index of the mobile phase must be compensated by means of a differential technique. Any substance whose refractive index differs sufficiently from that of the eluent can be detected. Accordingly, the differential refractometer has much greater universal application than a UV detector. However, this versatility also has some disadvantages. Thus, these instruments reveal every change in eluent composition and therefore cannot be used for gradient elution unless solvents are chosen with identical refractive indices. An additional disadvantage lies in the strong *temperature dependence* of the refractive index, which amounts to about 10^{-4} refractive index units (RIU) per ^{o}C. To attain adequate sensitivity (10^{-7} RIU), the temperature of the eluent and measuring cells must be held constant to $\pm$ 0.001^{o}C [3]. Consequently, this requires effective heat exchangers and a high heat capacity (e.g., a metal block) that buffers minute temperature fluctuations. Variations in the flow rate also interfere with the response of a differential refractometer. Very good damping is essential for pumps producing pulsating flows.

Commercial instruments operate on two different principles:

1. Fresnel-Refractometer

The basis of these refractometers is the Fresnel Reflection Law which states that the amount of light reflected from an interface depends on the angle of incidence (90-α) and the refractive index of both media forming the phase boundary.

Fig.IV.3 shows a schematic arrangement of such an instrument. The sample and reference cells are illuminated by the same lamp (B). Part of the incident light (R) is reflected at the interface between the liquid (F) and the prism surface. This part is not used for the measurement. The other part of the light penetrates the liquid layer and is diffusely backscattered by the rear steel plate (S) that bounds both cells and serves as heat exchanger. These portions of light pass through an optical system and strike separate photoresistors (D) for each cell. If the refractive index of the liquid flowing through the sample cell changes from that of the reference cell, a difference in brightness of the backscattered portions of light results. The differences in the photoresistances in then recorded. The volume of both cells between the prism (P) and steel plate (separated by a Teflon® mask (M)) can be kept very small (~5 μl).

The disadvantage of this arrangement lies in the inability to cover the entire refractive index region of LC solvents with a single prism. Prisms covering the ranges of 1.31 to 1.44 and 1.40 to 1.55 are offered with these instruments. Unfortunately, it is also difficult to maintain an adequate constancy of temperature between the eluent and the optical components, especially at higher flow rates (> 1 ml/min). Thermostating the steel plate alone is insufficient.

The detection limit of the Fresnel refractometer is $\pm 4 \cdot 10^{-7}$ RIU under optimum conditions. For example, $2 \cdot 10^{-6}$ g/g (corresponding to 2 ppm) of aniline in hexane can be detected.

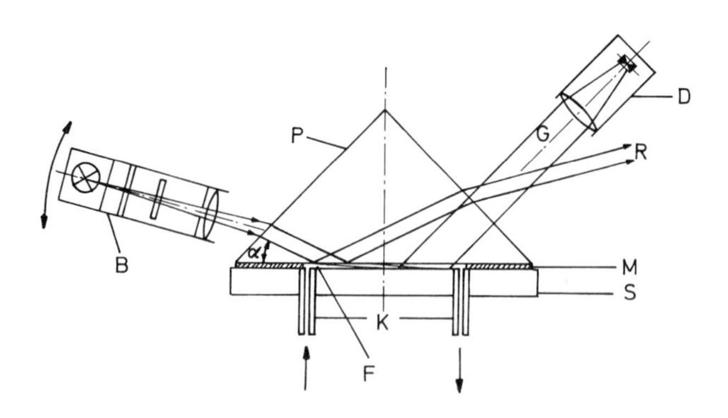

Fig.IV.3. Schematic of a fresnel refractometer (for explanation see text)

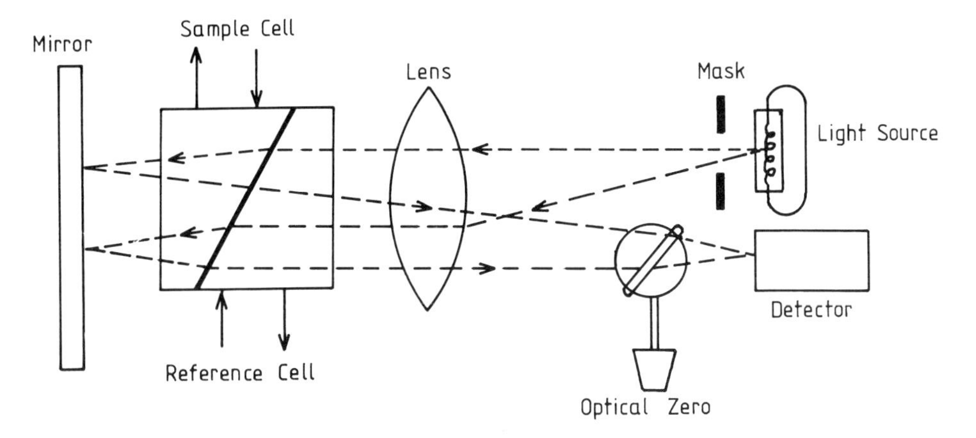

Fig.IV.4. Schematic of a deflection refractometer

2. Deflection Refractometer

If a light beam is passed through a cell filled with two liquids having different refractive indices, the deflection of the beam is proportional to the difference in the refractive indices. Fig.IV.4 shows a schematic of this type of refractometer. The light beam from the lamp is restricted by a mask, focused by a lens, and then strikes the cell, half of which is filled with the reference liquid and the other half with the column effluent. The beam is deflected, reflected by the mirror, passed once more through the cell where it is again deflected, and impinges on the detector, a photoresistor. If both cells contain a liquid with the same refractive index, the beam will impinge squarely upon the narrow slit of the detector. A change in refractive index in the sample cell causes the beam to be displaced, thus altering the point of impact and resulting in a weaker photocurrent. (At the start of the measurements the beam is adjusted by means of a rotatable glass plate to equalize the differences in the refractive indices).

With this instrument all possible refractive indices can be measured; changing the cell is unnecessary. The cell and attendent optical apparatus can be conveniently thermostated. It is best to install everything in a metal block, which also houses the heat exchanger [3]. The sample cell is double the size (10 μl) of a Fresnel-type refractometer. However, the sensitivity of a deflection refractometer is usually higher than that of the refractometers described above [4]. Under optimum conditions the noise level is less than $3 \cdot 10^{-8}$ RIU [8].

For the example of aniline in hexane this amounts to a theoretical detection limit of about 10^{-7} g/g, which corresponds to 0.1 ppm. Chromatographic concentrations in the ppm region can be routinely detected. Nevertheless, it must always be borne in mind that the concentration in the column effluent is always considerably lower than on injection. The dilution increases with stronger sample retention (i.e., higher k' values), greater plate height, and increased column length. For example, at about the same plate height for both compounds, 0.4 μg of Aldrin (k' = 0.2) and 2.2 μg of Endrin (k' = 9.2) could be detected by means of a deflection refractometer using a 10 μl cell [4].

A comparison of these values with those in Table IV.1 shows that UV detectors are generally more sensitive for the detection of compounds having molar absorptivities $\varepsilon > 10^2$ than the usual differential

Table IV.2. Refractive indices of important eluents at 20°C [5]

Acetone	1.3588	
Acetonitrile	1.3441	
Ethanol	1.3611	
Ethyl bromide	1.4239	
Benzene	1.5011	
Chloroform	1.4433	at 25°C
Cyclohexane	1.4266	at 19.5°C
Diethyl ether	1.3526	
Di-isopropyl ether	1.3679	
Dioxane	1.4224	
Ethyl acetate	1.3701	at 25°C
Methyl acetate	1.3617	
Heptane	1.38764	
Hexane	1.37486	
Methanol	1.3288	
Methylene chloride	1.3348	at 15°C
Pentane	1.3579	
i-Propyl chloride	1.3781	
n-Propyl chloride	1.3886	
Carbon tetrachloride	1.4664	
Tetrahydrofuran	1.4076	at 21°C
Toluene	1.4961	
Water	1.3330	

refractometers. The refractometers, however, respond to *all* components whose refractive index differs from that of the eluent, whereas UV detectors require the presence of a chromophore in the molecule.

Table IV.2 lists the most important eluents along with their indices of refraction. Chromatographically equivalent solvents, e.g., pentane, hexane, and heptane or chloroform and methylene chloride differ substantially in their refractive indices, thereby permitting improved detection sensitivities in critical cases by changing the eluent.

Other refractometers have been described for use in HPLC. One detector employs the Christiansen effect to measure changes in refractive index. In this case the detector cell contains a solid substance (e.g., glass) having the same refractive index as the mobile phase used. As long as the refractive indices of the solid and eluent remain equal, the cell is completely transparent to light. When the refractive index of the eluent is altered, as by the elution of a sample, the light transparency of the cell also changes. This change can be measured with a photocell. The sensitivity of this type of refractometer is supposedly of the same order of magnitude as that of the others. However, whether suitable solids with the proper refractive indices are available for every solvent and mixture is not known at this time.

Interferometers can also be used to monitor changes in the refractive index in the eluate. Because of the stringent requirements placed on the optical components of such instruments, they are quite expensive.

C. Microadsorption Detector

This detector is based on the measurement of the heat of sorption that is evolved with the passage of a sample band through a stationary phase [6]. Since sorption is always followed by desorption whereupon heat is absorbed from the eluent, the measurement should involve a positive response followed, in the ideal case, directly by an equal negative one. Theoretically, a differential Gaussian curve should be obtained. This detector is capable of responding to all substances, but, unfortunately, is still burdened with some systematic problems. A serious drawback lies in the unsymmetrical signal. The desorption signal is less sharp than the adsorption signal, and the area under the former is also frequently smaller. As a result, electronic integration fails to yield a Gaussian curve; the S-shaped signal obtained prevents identification of neighboring peaks. The exact determination of the retention times (elution of the center of mass) is essentially precluded. The height of the absorption peak is a function of the amount of sample, but it is readily affected by contamination or deactivation of the small amount of adsorbent in the sample cell. Moreover, the microadsorption detector is sensitive to fluctuations in the flow rate and even shows the pulsations of poorly damped piston pumps.

The measuring device is very simple. The thermistors can be mounted in the end of the column with hardly any dead volume. Unfortunately, it is difficult to compensate temperature fluctuations, so that a separate arrangement is preferred: Immediately after the thermistor embedded in the adsorbent, a second one mounted in the eluent stream and surrounded by glass beads compensates for the temperature fluctuations in the eluent and the ambient temperature. Even with this arrangement the dead volume remains minimal. Of course, thermostating to at least $\pm$ 0.003°C is necessary, otherwise a heat exchanger must be installed between the column and detector. This arrangement permits temperature changes of 10^{-4} $^{\circ}$C to be determined. At present, such a detector is not commercially available.

D. Transport Detector (Flame Ionization Detector)

This is the only detector where the eluent is removed before the sample is monitored. To do this, the sample is applied onto a transport device (chain, spiral, or wire). After evaporation of the eluent, the nonvolatile sample is conveyed to a flame ionization detector. The enormous advantage of this system, ideally, is its response solely to the sample. The response is therefore independent of the type of chromatographic development used. The chemical properties, temperature, and pulsations of the mobile phase do not affect sample detection. However, one condition must be fulfilled: There must be a very large difference in volatility between the eluent and sample so that enough of the sample residue remains on the transport system after evaporation of the eluent.

A schematic of a transport detector [8] is shown in Fig.IV.5.

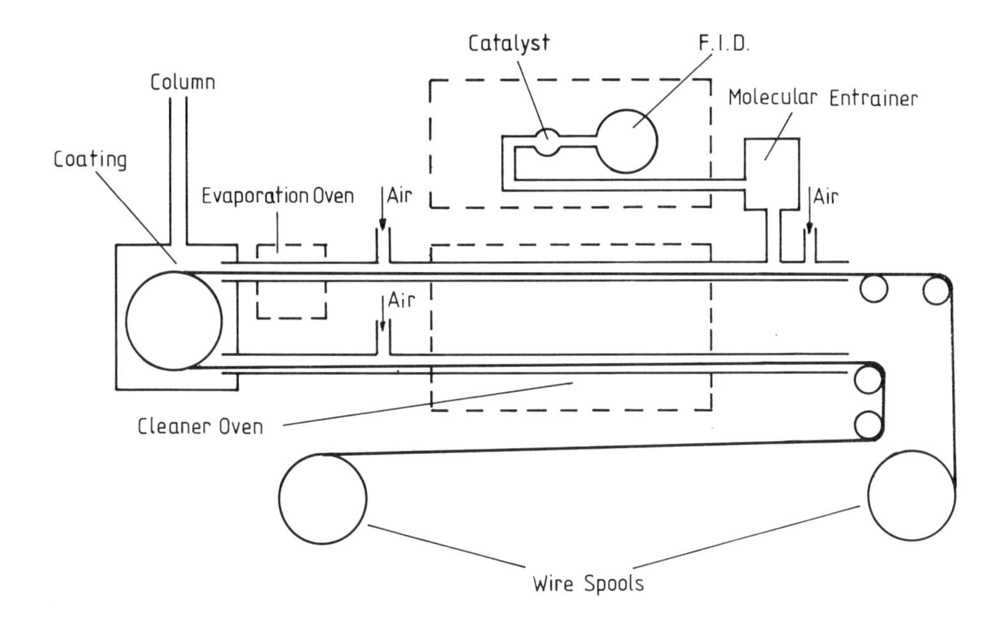

Fig.IV.5. Schematic of a wire transport detector (for explanation see text)

The stainless steel wire is cleaned by oxidation in a stream of air at high temperature and surface-coated with an oxide layer. (This treatment provides a surface amenable to very uniform coating by eluents, even those with a high surface tension). The wire is then fed through the eluent stream and coated with sample and eluent. The eluent is removed in an evaporation oven, whose temperature can be precisely regulated, and flushed out with air. The sample on the wire is then fed through an oxidation oven and burned at 600 to 800°C in a stream of air. The combustion gases are aspirated with a type of water aspirator pump driven by hydrogen (a molecular entrainer) and, following catalytic conversion to methane, sensed by a flame ionization detector (FID). The process of combustion and conversion to methane has advantages over the original pyrolysis because the combustion of many organic compounds proceeds more uniformly and reproducibly than their pyrolysis. The special wire used is 10 km long and 0.12 mm thick, and must, of course, be fed at a constant rate.

The sensitivity is not very high when compared to that of an FID in gas chromatography, but does fall into the concentration range of almost all LC detectors, namely 1 to 3 µg/ml (about 1 ppm). Considering the small amount of sample that adheres to the wire, the sensitivity is excellent. However, it must be remembered that a portion of the sample is lost during evaporation of solvent. The volatility differences of the sample components may lead to spurious quantitative results. Furthermore, not all of the combustion gases reach the FID. This detector should be ideal for HPLC because it would detect the sample irrespective of the eluent, and would permit gradient elution separations to be performed without any complications.

E. Fluorescence Detector

Substances that emit light in the visible region after being excited by UV radiation (usually in the near UV region) can be monitored by a fluorescence detector. The excitation wavelengths may either be predetermined by the source used (e.g., moderate- or high-pressure mercury lamp) or may be selected if a monochromator is employed. In most cases measurements are carried out perpendicular to the direction of excitation. Stray radiation from the excitation source must be removed by suitable filters.

The fluorescence detector is very specific and its use requires that more precautions be taken than, for example, with the UV detector. For example, the fluorescence may be quenched or suppressed by invisible contaminants. Some solvents (e.g., oxygen-containing compounds) quench the fluorescence and are therefore unsuitable. Obviously, eluents that absorb in the region of excitation cannot be used.

In many cases the sensitivity of a fluorescence detector exceeds that of a UV detector. Thus, 10^{-9} g/ml or 1 ppb of quinine sulfate, a favorite standard in fluorometry, can be detected. The region of linearity of fluorescence detectors is greater than that of UV instruments and often exceeds five orders of magnitude.

Many classes of compounds can be converted to fluorescent derivatives, which considerably enhances their detection sensitivity. Thus, amino acids, alkaloids, and catecholamines can be converted to dansyl derivatives (with 1-dimethylaminonaphthalene-8-sulfonic acid) and detected as low as the nanogram range [27,28,29]. The separation of the dansyl derivatives can be accomplished without great difficulty, even though the molecular properties become very similar as a result of the derivatization.

F. Other Detectors

1. Electrochemical Detectors

The application of electrochemical redox reactions (polarographic and amperometric methods) to sample detection in the eluate has also been described [9,10,23,30,31,52-58]. In general, a constant voltage is applied to a pair of electrodes and the current resulting when a sample is oxidized or reduced is measured as a function of time. Practically, the oxidation method can be carried out more conveniently because the eluent can be used without pretreatment, i.e., the removal of dissolved oxygen, which is essential prior to the utilization of the reduction method. The oxidation or reduction voltage and hence the possibility of detection depends on the oxidation or reduction potential of the solvent. Of course, the eluent must possess a certain electrical conductance. For aqeous systems 0.05 M KNO_3 is satisfactory, whereas for organic eluents tetraethylammonium perchlorate (0.05 M) is recommended. A potential of 1 volt (*vs.*S.C.E.) can be applied in aqueous and aqueous alcohol solutions. At this potential organic nitrogen compounds (e.g., amines, amino acids, heterocyclics, etc.), nitro compounds, phenols, aldehydes, and ketones can be detected. Since the decomposition potential differs for each class of compounds, the detector can be adjusted to respond very selectively to a particular class by varying the applied potential or to detect all classes by choosing an upper value.

In addition to the classical dropping mercury electrode [9], special electrodes such as the graphite-impregnated silicone rubber electrode [10] or pure carbon electrode [30] are used. Solid electrodes must be freed of the oxidation products. This can be accomplished most elegantly by a pulse technique in which, in addition to the positive working potential, a brief negative cleaning (reduction) potential is applied. Spraying the electrode with eluent [30] enhances this cleaning step.

The sensitivity of this detector depends strongly on the ease of oxidation or reduction of the sample. Presently available data suggest great promise for this device as it can be used to detect important substances such as adrenalin and its derivatives in biologically important concentration ranges.

2. Conductivity Detector

The conductivity detector is very specific in its response because only ions can be detected. This restricts its utility to water and polar eluents. Commercial instruments can be used in both the absolute and differential mode, the precision being greater for relative measurements than absolute ones. To achieve adequate sensitivity in eluents with high conductances, the instruments are equipped with zero suppression.

The measurement of the conductivity itself presents certain problems: If a direct current source is used, polarization effects interfere with the conductivity at the higher ionic concentrations. An alternating current source extensively suppresses the polarization, but this same system also measures changes in the dielectric constant, which swamp the conductivity measurement, especially at low ionic concentrations. Instruments with a.c. sources should therefore be equipped with a detector that measures only on the ascending phase of the a.c. cycle, thereby suppressing the interference caused by the dielectric constant that appears on the descending phase.

The cell volume can be kept very small - about 2 µl. For automatic temperature compensation the cells are equipped with a thermistor. Good temperature constancy should be provided since the conductivity of a solution may change by about 2% per $^{\circ}C$. It is advantageous to calibrate in terms of the specific conductance as this eliminates the need to determine the cell constant. The calibration must be repeated, however, when another cell is used. These instruments are so sensitive that conductivity differences of $5 \cdot 10^{-4}$ µmho/cm can be determined. Such changes are caused by only 5 to 10 ppb of sodium chloride in water. In buffer solutions the sensitivity is, of course, smaller because of their own conductance. The linear region extends from about 0.01 to 100 000 µmho/cm.

3. Capacity Detector

Capacity measurements have also been used to monitor sample concentrations in the eluent. Large changes in the dielectric constant, and, hence, a highly sensitive response is to be expected when the dielectric constants of the sample and eluent differ greatly. To compensate for temperature effects, a differential technique with a reference cell may be employed. No dielectric constant detector specifically for

HPLC is yet commercially available, although various designs for one have been described [11-13] and the theoretical possibilities discussed [14,15]. The cells must be constructed so that the distance between the electrodes remains very stable. Despite the relatively large surface area required for both electrodes, the cell volume can be kept very small. For a given noise level, these detectors are said to be more sensitive than a differential refractometer. The sensitivity is independent of the flow rate. The sensitivity is quoted as being 0.9 ppm for chloroform in isooctane ($\Delta\varepsilon \sim 3$)[13], 0.4 ppm for acetone in n-hexane ($\Delta\varepsilon = 19.9$), and 260 ppm for n-octane in n-hexane ($\Delta\varepsilon = 0.06$).

4. Radioactivity Detectors

Radioactivity measurements are only feasible in the separation of adequately labeled compounds. Detection is very specific and is not affected by changes in the eluent composition. There are special scintillation counters for HPLC with a cell volume of 200 μl. However, ordinary commercial instruments with large dead volumes (~ 1 ml) can also be employed, provided the band broadening caused by this dead volume can be tolerated [60]. In the procedures described, the scintillators are brought into direct contact with the eluent [16,17]. Sometimes the eluent flow can be stopped when the majority of the labeled substance is in the sample cell in order to increase the counts from weak emitters. Of course, the eluent must be transparent to the scintillation radiation. Impurities, pressure on the cell, and variation in the flow rate, as well as factors affecting the stationary count, interfere with the response of the detector.

5. Directly Interfaced HPLC-Mass Spectrometry

The direct interfacing of a liquid chromatograph with a mass spectrometer presents far more problems than a GC-MS coupling. Two basic LC characteristics impede direct interfacing. The amount of eluent is so large that it overwhelms the usual vacuum system of a mass spectrometer. At an eluent flow rate of 1 ml/min between 150 and 1200 ml/min of vapor is normally formed, whereas a modern vacuum system will normally handle 1 to 20 ml/min maximally if it is set up for chemical ionization. Furthermore, the greatest field of application of HPLC lies in the separation of nonvolatile or slightly volatile samples. Thus, the second problem entails sample vaporization without decomposition, which can be solved more readily, however, because of the relatively short residence time of the sample in the vaporization zone.

Several types of interfaces have already been utilized:

1. Vaporization of eluent and sample at atmospheric pressure and ionization under these conditions [21,22]. Only a fraction of the ions is admitted into the vacuum system through the slit. Mass spectra can be recorded in the nanogram sample range.

2. Splitting of the effluent stream so that about 10 μl reaches the ion source of the mass spectrometer and is vaporized there [25,32]. If chemical ionization is used, the relatively high solvent flow does not interfere, although the eluent must have compatible properties. Ions with higher mass numbers than the molecular weight of the sample are produced because the eluent molecules (or their fragments) may attach themselves to the molecular ion (or its fragment ions). The explanation and interpretation of such spectra thus can become very difficult.

3. The use of enrichment systems involving preseparation of sample and eluent appears to be more promising. The most obvious means of conveying the sample to the mass spectrometer following removal of the eluent seemed to be either to use the transport wire detector [24] (cf. Section D) or a ribbon with a large sample capacity [33]. Using the transport mechanism in conjunction with an eluent having a boiling point below $80^{\circ}C$, only about 10^{-7} g/sec of the non-vaporized eluent reaches the mass spectrometer, whereas between 20 and 40% (of 10^{-8} g) nonvolatile sample (such as methyl stearate) is conveyed to the mass spectrometer [33]. The maximum tolerable eluent flow rate varies between 2 ml/min for hexane and 0.2 ml/min for water. At higher flow rates more eluent is transported into the mass spectrometer. The sensitivity limit in determining the total ion current corresponds to 10^{-6} g/ml, i.e., about 1 ng is available for obtaining the mass spectrum, a quantity that is generally sufficient. Membrane enrichment systems between the column outlet and the MS inlet have also been utilized [34]. Field desorption MS, in which only few fragment ions are produced in addition to the molecular ion, may offer further advantages for LC-MS interfacing [18].

6. Infra-red Detector

A commercial IR detector has recently become available. Most of the common eluents absorb in the IR and "windows" exist only at a limited number of wavelength regions where samples may be monitored by their functional group absorbances. The most important window lies in the carbonyl region (1600 - 1800 cm^{-1}) where most eluents are transparent. Due to their high absorptivity in this region, carbonyl-containing samples can be detected with little difficulty. The C-H stretch vibrations at 3000 cm^{-1} may be used similarly with suitable eluents (e.g., CCl_4).

Fats have been monitored by means of this detector from a methylene chloride-acetonitrile eluent [62]. It has also been used for the characterization of polymers. However, because of the relatively weak absorptivity of IR bands the potential for application of this detector appears to be restricted to the selective detection of carbonyl groups and, in favorable cases, C-H vibrations in conjunction with C-H-free eluents.

7. Other Methods

In addition to the above devices, those based on changes in the density, viscosity, heat capacity, vapor pressure, and sonic velocity have been utilized to determine the sample concentration in the eluent. A mass detector has also been described [20] in which the amount of sample residue remaining after solvent evaporation is determined with a micro balance and recorded continuously. Recently, the spray impact detector [61] has been described, whose response appears to be quite general, but whose sensitivity is rather variable and is markedly affected by the solvent used. Nevertheless, it would seem to merit further investigation.

All such detectors (as well as an ionization detector [26]) are in early stages of development and most are not commercially available. Hence nothing precise can be stated here concerning their applicability and sensitivity.

The use of the electron capture detector (ECD), which has extensive application in GC pesticide analysis, for the detection of halogen-containing substances in a column effluent has been described [35] and its application to the analysis of milk residues has been discussed [36]. In this case the entire column effluent is vaporized, flushed with a purge gas (e.g., nitrogen) to the ECD, and the vapors are then recondensed. Of course, the eluent and its impurities should not exhibit an electron capture reaction. Up to now, only separations with n-hexane have been described, to which up to a maximum of 5% of aliphatic alcohol or ether had been added.

For the detection of inorganic compounds an atomic adsorption spectrometer interfaced with a liquid chromatograph has also been described [59].

Polymeric samples have been monitored by means of light scattering in exclusion chromatography.

G. Comparison of the Important Detectors

Some important specifications for detectors are summarized in Table IV.3. Only three of the commercially available detectors have general applicability, i.e., they respond more or less to all substances. Unfortunately, they are not the most sensitive ones, and, furthermore, their responses are strongly temperature dependent.

Table IV.3

Detector	Generally applicable?	Physical units	Max. sensitivity (most favorable case)	Temperature dependence
UV detector	no	AU	10^{-10} g/ml	negligible
Refractometer	yes	RIU	10^{-7} g/ml	10^{-4} RIU/^{o}C
Adsorption	yes	^{o}C	10^{-9} g/sec	$5 \cdot 10^{-5}$ ^{o}C
Transport	yes	amp	10^{-7} g/ml	none
Conductance	no	mho/cm	10^{-8} g/ml	2%/^{o}C
Fluorescence	no	AU	10^{-9} g/ml	none
Capacity	yes	F	10^{-7} g/ml	10^{-3}/^{o}C
Polarographic	no	amp	10^{-9} g/ml	1.5%/^{o}C

The maximum sensitivities quoted have been taken from the manufacturers' literature and generally represent statically measured values in which the stated concentrations were placed directly into the sample cell.

The cell volumes of all detectors range between 2 and 10 μl, thus largely conforming to the needs of the practitioner. For some designs the noise level and hence the detector sensitivity depend on the flow rate.

The linear range for most detectors is 1:1000 and should be adequate for quantitative analysis. For the UV detector, the more nearly monochromatic the UV light, the greater is the linear range.

The response of almost all the detectors described here is concentration specific, that is, the magnitude of the signal is independent of the flow rate. Only the micro adsorption detector is sensitive to the magnitude of flow, the signal increasing with rising flow rate. In GC, the FID is a flow-dependent detector; in LC, however, it is used in conjunction with a transport system that conveys the sample to the FID in amounts depending on its concentration in the column effluent. Therefore, in this case the FID is also concentration specific.

H. Reaction Detectors

The detector sensitivity for many important substances such as amino acids is presently too low to sense them at their naturally occuring concentrations. Therefore, to increase the sensitivity recourse must be had to the preparation of derivatives that elicit a greater detector response. In the case mentioned, dansyl derivatives can be prepared which are more readily monitored with a UV or fluorescence detector than the aliphatic amino acids with a refractometer [27,37]. Of course, all the derivatives employed in organic analytical chemistry may be used, e.g., phenylisocyanates [38] or anilides [39] of fatty acids, nitrobenzoates of sugars and alcohols [40], etc.

The detection sensitivity may be increased by allowing a chemical reaction to occur between the column outlet and the detector. In doing so, however, some basic difficulties are encountered. The reaction should not ruin the separation of the sample bands achieved in the column. Yet, good mixing of the reagent and effluent is essential.

Furthermore, since the reaction is carried out in a flowing medium and requires a certain amount of time for completion, the mixture must traverse a certain distance or volume before being swept into the photometric detector. The additional band spreading that accrues in this reaction volume must be kept negligibly small.

The Technicon principle, where the eluent is divided into segments by means of air bubbles, cannot be applied in this case because of the large band broadening that would ensue in HPLC. However, by electronic suppression of the effect of air bubbles on the detector, a successful chromatogram can be obtained [48]. Also, with very selective identification reactions these systems can be used for the specific determination of, for example, organophosphates by the cholinesterase method [41].

Fast reactions, such as the detection of amino acids or amines with Fluram®, which rapidly produces a short-lived fluorescence in the visible region and requires no lengthy reaction distance, have found extensive application despite the relatively high cost of the reagent.

Suitably *deformed tubes* may be used to carry out the classical ninhydrin reaction without additional band spreading [42]. The detection limit for separations using columns packed with 10 μm particles is 0.1 - 1 ng. Other reactions [42] can also be performed via this generally applicable method of deformed tubes [43].

Very efficient *packed columns* are also suitable as a reactor for reactions performed after a separation. However, the pressure drop of the total system (separation and reaction columns) exceeds that of the deformed tube design. For this system reactions yielding fluorescent or colored products have also been investigated [46,49], a Ninhydrin detector [63] has also been reported. A reaction detector has been described for the detection of sugars in the eluate of ion exchangers (borate complexes) [47].

Treatment of the eluate with Ce(IV) ions, which are reduced to fluorescent Ce(III) ions, has also been employed for the detection of samples after separation [44,45].

Serious fundamental difficulties are associated with these reaction detectors. A detector must be designed and optimized for each individual reaction. Generally applicable reaction detectors may not be commercially available in the immediate future but rather, complete systems for a specific separation problem (such as sugar or amino acids analysis) may be developed.

References Chapter IV

Review

Byrne, S.H., jr., in: Kirkland, J.J. (Ed.): Practice of High Speed Liquid Chromatography. New York: Wiley-Interscience 1971

Polesek, J., Howery, O.G.: J. Chromatogr. Sci. *11*, 226 (1973)

1. Felton, H.: J. Chromatogr. Sci. *7*, 13 (1969)
2. Williams, D.H., Fleming, I.: Spektroskopische Methoden in der organischen Chemie. Stuttgart: Thieme-Verlag 1968
3. Deininger, G., Halâsz, I.: J. Chromatogr. Sci. *8*, 499 (1970)
4. Bombaugh, K.J., Levangie, R.F., King, R.N., Abrahams, L.: J. Chromatogr. Sci.*8* , 657 (1970)
5. Handbook of Chemistry and Physics. 46th Edition. Chemical Rubber Co. 1965
6. Hupe, K.-P., Bayer, E.: J. Chromatogr. Sci. *5*, 197 (1967)
7. Lapidus, B.M., Karmen, A.: J. Chromatogr. Sci. *10*, 103 (1972)
8. Scott, R.P.W., Lawrence, J.G.: J. Chromatogr. Sci. *8*, 65 (1970)
9. Koen, J.G., Huber, J.F.K., Poppe, H., Den Boef, G.: J. Chromatogr. Sci. *8*, 192 (1970)
10. Joynes, P.L., Maggs, R.S.: J. Chromatogr. Sci. *8*, 427 (1970)
11. Haderka, S.: J. Chromatogr. *52*, 213 (1970)
12. Vespalec, R., Hana, K.: J. Chromatogr. *65*, 53 (1972)
13. Poppe, H., Kuysten, J.: J. Chromatogr. Sci. *10*, 16 A (1972)
14. Haderka, S.: J. Chromatogr. *54*, 357 (1971)
15. Haderka, S.: J. Chromatogr. *57*, 181 (1971)
16. Schram, E., in: Bransome, E.D., jr. (Hrsg.): Current Status of Liquid Scintillation. New York: Grune and Stratton 1970
17. Hant, J.A.: Anal. Biochem. *23*, 289 (1968)
18. Schulten, H.R., Beckey, H.H.: J. Chromatogr. *83*, 315 (1973)
19. Carr, D.: Varian Instrument Applications *7*, 14 (1973)
20. Schulz, W.W., King, W.H., jr.: J. Chromatogr. Sci. *11*, 343 (1973)
21. Horning, E.C., Horning, M.G., Carrol, D.J., Dzidic, J., Stillwell, R.N.: Anal. Chem. *45*, 936 (1974)
22. Horning, E.C., Carrol, D.J., Dzidic, J., Haegele, K.D., Horning, M.G., Stillwell, R.N.: J. Chromatogr. Sci. *12*, 725 (1974)
23. Davenport, R.J., Johnson, D.C.: Anal. Chem. *46*, 1971 (1974)
24. Scott, R.P.W., Scott, C.G., Munroe, M., Hess, J., jr.: J. Chromatogr. *99*, 395 (1974)
25. Arpino, P.J., Dawkin, B.G., McLafferty, F.M.: J. Chromatogr. Sci. *12*, 574 (1974)
26. Mowery, R.A., jr., Juvet, R.S., jr.: J. Chromatogr. Sci. *12*, 687 (1974)
27. Bayer, E., Grom, E., Kaltenegger, B., Uhmann, R.: Anal. Chem. *48*, 1106, (1976)
28. Frei, R.W., Santi, W., Thomas, M.: J. Chromatogr. *116*, 365 (1976)
29. Schwedt, G., Bussemas, H.H.: Chromatographia *9*, 17 (1976)

30. Application sheet, Edt. Research, 65 Ivy Crescent, London W 4

31. Blank, C.L.: J. Chromatogr. *117*, 35 (1976)

32. Baldwin, M.A., McLafferty, F.W.: Org. Mass Spectrom. *7*, 1111 (1973)

33. McFadden, W.H., Schwartz, H.L., Evans, S.: J. Chromatogr. *122*, 389 (1976)

34. Jones, P.R., Yang, S.K.: Anal. Chem. *47*, 1000 (1975)

35. Willmott, F.W., Dolphin, R.J.: J. Chromatogr. Sci. *12*, 695 (1974)

36. Dolphin, R.J., Willmott, F.W., Mills, A.D., Hoogeveen, L.P.: J. Chromatogr. *122*, 259 (1976)

37. Lawrence, J.F., Frei, R.W.: Chemical derivatization in liquid chromatography. Amsterdam: Elsevier 1976

38. Borch, R.F.: Anal. Chem. *47*, 2437 (1975)

39. Hoffmann, N.E., Lino, J.C.: Anal. Chem. *48*, 1104 (1976)

40. Nachtmann, F., Spitzy, H., Frei, R.W.: J. Chromatogr. *122*, 293 (1976)

41. Ramsteiner, K.A., Hormann, W.R.: J. Chromatogr. *104*, 438 (1975)

42. Dissertation U.Neue, Saarbrücken 1976

43. Halász, I., Walkling, P.: Ber. Bunsenges. *74*, 66 (1970)

44. Katz, S., Pitt, W.W., jr., Mrochetz, J.E., Dinsmore, S.: J. Chromatogr. *101*, 193 (1974)

45. Katz, S., Pitt, W.W., jr., Mrochetz, J.E.: J. Chromatogr. *104*, 303 (1975)

46. Muusze, R.G., Huber, J.F.K.: J. Chromatogr. Sci. *12*, 779 (1974)

47. Zech, K., Voelter, W.: Chromatographia *8*, 7, 350 (1975)

48. Snyder, L.R.: J. Chromatogr. *125*, 287 (1976)

49. Deelder, R.S., Kroll, M.G.F., van den Berg, J.H.M.: J. Chromatogr. *125*, 307 (1976)

50. Denton, M.S., De Angelis, T.P., Yacynych, A.M., Heinemann, W.R., Gilbert, T.W.: Anal. Chem. *48*, 20 (1976)

51. Milano, M.J., Lam, S., Grushka, E.: J. Chromatogr. *125*, 315 (1976)

52. Kissinger, P.T., Refshange, C., Dreiling, R., Adams, R.N.: Anal. Lett. *6*, 465 (1973)

53. Kissinger, P.T., Felice, L.J., Riggin, R.M., Pachla, L.A., Wenke, D.C.: Clin. Chem. *20*, 992 (1974)

54. Riggin, R.M., Rau, L., Alcorn, R.L., Kissinger, P.T.: Anal. Lett. *7*, 791 (1974)

55. Riggin, R.M., Schmidt, A.L., Kissinger, P.T.: J. Pharm. Sci. *64*, 680 (1975)

56. Buchta, R.C., Papa, L.J.: J. Chromatogr. Sci. *14*, 213 (1976)

57. Tjaden, U.R., Lankelma, J., Poppe, H.: J. Chromatogr. *125*, 275 (1976)

58. Lankelma, J., Poppe, H.: J. Chromatogr. *125*, 375 (1976)

59. Jones IV, D.R., Tung, H.C., Manahan, S.E.: Anal. Chem. *48*, 7 (1976)

60. Figge, K., Piater, H., Kolbe, W.: G-I-T, Fachz. Lab. *19*, 192 (1975)

61. Mowery, R.A., Juvet, R.S., jr.: J. Chromatogr. Sci. *12*, 687 (1974)

62. Parris, N.A.: Abstracts 29th Pittsburgh Conference 1978, Paper No. 122

63. Jonker, K.M., Poppe, H., Huber, J.F.K.: Chromatographia *11*, 123 (1978)

Chapter V

Stationary Phases

The packing materials for HPLC columns must be pressure-stable. The permeability of a packed column should never be a function of the pressure drop along the column. The majority of inorganic supports are stable up to about 600 atm, although the pore structure of highly porous materials (pore volume > 2 ml/g) may collapse even at lower pressures.

Materials that swell, such as ion exchangers with an organic matrix, exhibit diminishing, i.e., poorer, column permeability with increasing pressure. This deterioration may proceed to the extent that the eluent velocity starts to decrease. However, some fairly pressure-stable purely organic ion exchangers are available (Section C).

In HPLC two different types of supports are used:

1. Totally *porous materials* (such as silica gel and alumina) with a large specific surface area (50 - 500 m^2/g) and high pore volume (0.2 - 2 ml/g) as employed in classical column chromatography, but with particle sizes less than 50 µm.

2. *Porous layer beads (PLB)*, where a thin, porous, active layer is coated onto a solid core, such as an impervious glass bead. The thickness of the porous layer is generally 1 - 3 µm (1/30 - 1/40 of the particle diameter). The glass beads are coated with silica gel and alumina, as well as ion exchange resins, polyamide, etc. Their particle size lies between 25 and 50 µm, and they produce greater column efficiency than porous particles of the same diameter (cf. Table III.1). However, the importance of PLB has diminished somewhat since the development of good and reproducible packing techniques for 10 or 5 µm diameter porous particles.

The particle size distribution (sieve fraction) of all packing materials should be as narrow as possible. Particles greater than 20 µm can be easily classified by sieving, wet-sieving being preferred for < 40 µm diameter particles. For particles less than 20 µm in diameter recourse to cyclone sieving is necessary, and for those with d_p < 10 µm only sedimentation can yield narrow fractions. One should make it a rule to remove fines (dust, etc.) from a stationary phase prior to use by sedimentation in water (methanol or acetone for very small particles). Such treatment primarily improves column permeability, and helps prevent fine particles from clogging the connecting tubing to the detector. The better, i.e., narrower the sieve fraction, the easier it appears to be to pack efficient columns. To illustrate this, similar columns were packed with two nominal 7 μm silica gel fractions, one containing particles ranging between 5.6 and 8.4 μm and the other between 2 and 20 μm [1]. The d_{50} value (median value) of both fractions was 6.8 μm. At flow rates of 2 ml/min the column containing the narrower fraction exhibited an H value of 29 μm and a pressure drop of 72 atm, whereas the corresponding values for the broader fraction were rather high - 450 μm and 370 atm. Thus, for relatively broad fractions the worst possible case always seems to operate: the smallest particles of a sieve fraction exert the greatest effect on column permeability whereas the largest affect column efficiency.

A. Packing Materials for Adsorption and Partition Chromatography

In principle, either PLB or totally porous materials with a specific surface area greater than 2 m^2/g are applicable to adsorption and partition chromatography. PLB coated with a thin layer of silica are sold under different names (Actichrom, Pellosil, Perisorb, Vydac, Zipax, etc.). These materials are pressure-stable, but the amount of active stationary phase per unit volume is small compared with totally porous materials. The specific surface area of PLB coated with silica amounts to only 0.2 - 15 m^2/g. However, the actual surface area of the silica layer is considerably greater and corresponds to that of porous particles [2]. Some PLB materials, such as Zipax®, possess such a small specific surface area that they can only be used after being coated with a liquid stationary phase. These materials can be coated, before being packed into a column, with about 1 - 2% w/w of a liquid phase without becoming sticky. PLB containing an alumina layer or polyamide film are also available. PLB serve advantageously for rapid separations or in cases where the components are excessively retarded on porous materials.

Of the numerous totally porous sorbents employed in classical column chromatography, silica gel and alumina are by and large the only polar ones used for HPLC. Silica gel is sold as such or under a variety of brand names (Lichrosorb, Nucleosil, Partisil, Spherosil, Spherisorb, Zorbax, etc.). It should be borne in mind that these stationary phases are more or less amorphous gels whose properties not only vary from brand to brand but also from batch to batch and may additionally change during storage, purification, heating, reaction, etc. Especially large differences can be expected between separations performed on irregular silica (prepared from sodium silicate) and spherical silica (usually prepared in the presence of organic compounds) because of the differences in their surface properties, particularly the silanol concentration, (cf. Chapter V.I.B). Owing to their importance, the properties of silica gel and alumina will now be discussed in greater detail.

1. Silica Gel

Silica gel is the most frequently used adsorbent. It often also serves as a support for liquid stationary phases in partition chromatography (cf. Chapter VII). Silica gels, including that of the occasionally used porous glasses, are amorphous and can be prepared [3,4] in high purity and with various physical properties (specific surface area, pore volume, and pore size). For chromatographic purposes, primarily those with a relatively large specific surface area (> 200 m^2/g), large pore volume (> 0.7 ml/g) and moderate pore diameter (80 - 150 Å) are useful.

The effect of the pore structure of silica gel on a chromatographic separation is shown in Fig.V.1. The separation of oligophenyls was carried out under identical conditions (n-heptane, moisture 20% of saturation) on silica gel having different pore structures [5]. The specific surface area decreases from 250 m^2/g for SI 100 to 6 m^2/g for SI 4000. The optimum separation is attained on a silica gel with an area of 50 m^2/g (SI 500). The absolute retentions are a function of the specific surface area, but the relative retentions should be independent provided the surface structure does not change. For smaller pore diameters (< 60 Å) than those used here, an exclusion effect may be observed even for relatively small molecules.

The surface hydroxyl groups (silanol groups) may be arranged so that they are isolated or are capable of forming hydrogen bonds with adjacent OH groups [6]. Water is adsorbed by hydrogen bonding. Thermal

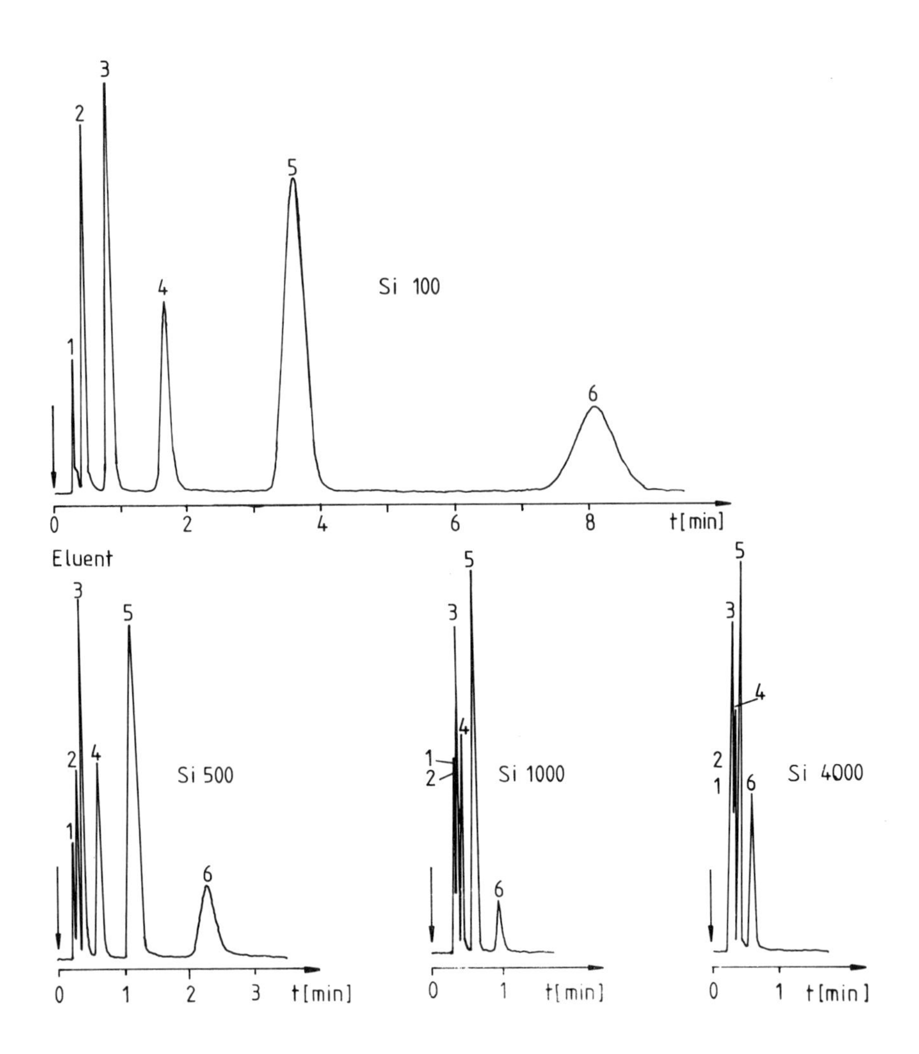

Fig.V.1. Effect of silica pore structure on separation (Merck Application 75 - 36). Stationary phase: Lichrospher Si 100, Si 500, Si 1000, Si 4000; d_p ~ 10 µm; column length: 20 cm, 3 mm i.d.; eluent: n-heptane (20% rel. moisture); F = 5 ml/min; Δp = 125 atm; samples: 1 = benzene; 2 = diphenyl; 3 = m-terphenyl; 4 = m-quaterphenyl; 5 = m-quinquephenyl; 6 = m-sexiphenyl [5]

treatment below 150°C removes only the physically adsorbed water. Surface reactions occur above 200°C. Neighboring hydroxyl groups condense between 300 and 500°C to form siloxane groups with the elimination of water. At still higher temperatures even the isolated groups, the so-called free hydroxyl groups, are split off. Silica gel heated in this manner is no longer suitable for adsorption chromatography. That which has been dehydrated at high temperatures is hydrophobic and is no longer selective in its adsorption of polar molecules. Water added to partially dehydrated silica is only physically adsorbed. At room temperature scarcely any rehydration of siloxane groups occurs.

The data on the number of hydroxyl groups per unit of surface area vary widely, depending on the history of the silica and the method of their determination [3]. A good value appears to be 5 hydroxyl groups/100 Å^2, corresponding to about 8 μmol OH/m^2 of surface. Unsaturated and polar molecules are adsorbed almost exclusively on these surface hydroxyl groups.

By chemical modification of the surface silanols, hydrophobic stationary phases with reversed-phase properties are obtained. If these derivatives contain functional groups, the properties and selectivity of the silica gel are modified accordingly [8-20].

Because the silica gel surface is weakly acidic, basic substances, especially in polar eluents, are more strongly retained than acidic or neutral substances. Commercial products are frequently neutralized by the addition of bases. In polar media, particularly in water, the acid reaction of the silica gel surface may interfere, but chemical changes during separation are relatively rare [21].

2. Alumina

Alumina suitable for chromatography is prepared by dehydration of bayerite, followed by activation at 200 - 500°C. The alumina obtained is crystalline (γ-Al_2O_3) and changes on heating to 900 - 1000°C into high temperature forms that convert at 1100°C to α-alumina, which no longer possesses any chromatographic activity ("dead-burnt" Al_2O_3).

The surface area of alumina lies between 100 and 200 m^2/g, with a pore volume of 0.2 - 0.3 ml/g. The average pore diameter is between 100 and 200 Å. Aluminas prepared at higher temperatures have a smaller specific surface area (70 - 90 m^2/g).

The sorption mechanism on alumina is more complex than that on silica gel [6]. In addition to the formation of hydrogen bonds to the surface hydroxyl groups or oxygen atoms, there is also the possibility of interactions of basic (electron-rich) molecules with the Lewis acid sites on aluminum atoms. Adsorption on these Lewis acid sites is partially irreversible. Coating with water leads to the loss of this activity.

The strongly active Lewis acid sites cause decomposition of sensitive substances during separation. The addition of 1 - 3 weight % of water to the active alumina eliminates this effect [6,22]. Another reason for the decomposition reaction on alumina may be its basic or acidic surface reactions. In the preparation of alumina the residual alkali content of bayerite is bound as sodium aluminate to the surface. An aqueous suspension of such alumina has a pH of 9 and is therefore designated as "basic". The surface sodium aluminate sites can act as cation exchangers in polar media, especially in water. This may lead to irreversible adsorption of cationic compounds or the decomposition of alkali-sensitive substances. Treatment of basic alumina with strong acids such as HCl results in an exchange:

$$>AlO^{\ominus}Na^{\oplus} + 2HCl \rightarrow >AlCl + NaCl + H_2O \quad .$$

The treated alumina has a pH of 3 in aqueous suspension and is therefore referred to as "acidic" alumina; it is an anion exchanger. By careful neutralization a "neutral" alumina is obtained, which is neither a cation nor an anion exchanger. Its aqueous suspension has a pH of 6.8 and it does not cause interferences via basic or acidic surface reactions. "Neutral" alumina should not retain either methylene blue (absorbed by "basic" cationic sites) or naphthol orange (adsorbed by "acidic" anionic sites) from aqueous solution. "Neutral" alumina is preferred for the chromatography of sensitive substances even in nonpolar media because water is almost always adsorbed on its surface, in which the retarded components dissolve and thereby come in contact with the acidic or basic surface in a polar medium.

Alumina belongs to the polar adsorbents, and its separation properties closely resemble those of silica gel. However, unsaturated molecules are generally more strongly retained on alumina. Condensed aromatics can be separated more readily on alumina than on silica. Interferences as a result of irreversible adsorption ("chemisorption") or the surface reactions discussed are more prevalent on alumina.

3. Polyamides

Polyamides are suitable for the separation of compounds capable of forming hydrogen bonds (phenols, quinones, sugars, etc.). The extent of sorption depends on the number of peptide groups in the polymer (whether Nylon 4, Nylon 6, or Nylon 11). Sometimes it is necessary to acetylate the free amino end groups of the polyamide in order to avoid irreversible adsorption.

A pure polyamide is unsuitable for HPLC. However, it can be very easily precipitated onto nonporous as well as porous supports, and such phases are commercially available.

Sorption on polyamide is strongest from solvents incapable of forming hydrogen bonds. Successful elution can be obtained with solvents such as alcohols, water, or dimethylformamide. The sorbed material can be completely displaced with sodium hydroxide.

Silica and alumina can be used for adsorption chromatography (Chapter VI) and become suitable for partition chromatography (Chapter VII) after being coated with a liquid phase. The average pore diameter of the support seems to have no effect on the chromatographic behavior if it is greater than 60 Å and less than 700 Å. The smaller the specific surface area of a support, the smaller are the k' values for a given sample, assuming the same eluent composition. The maximum coverage of liquid phase for partition chromatographic applications corresponds approximately to the pore volume of the solid (e.g., silica gel with a pore volume of 1 ml/g can be coated with about 1 g of liquid phase per gram of silica. In chromatography this is referred to as a "100% coverage"). [A detailed discussion of the properties of stationary phases is presented in Chapters VI and VII.]

B. Chemically Modified Supports

The surface hydroxyl groups of silica and alumina govern the adsorption properties and selectivity of the stationary phase. By bonding organic compounds to these hydroxyl groups, the chromatographic behavior of the solid can be altered drastically. The structure of the solid (i.e., the specific surface area, pore diameter, etc.) determines the properties of the chemically bonded stationary phase. Silica gel is used almost exclusively for the preparation of chemically modified supports. If the organic residue exerts no specific or selective effect, the k'values on such supports are always lower than on bare silica gel, assuming the same eluent composition. The selectivity of the stationary phases can be modified by varying the organic component, for example, by introducing functional groups into the organic residue, preferably in the ω-position. Since it is impossible to react *all* the surface hydroxyl (silanol) groups on the silica, the selectivity of the stationary phase is also affected by the remaining ones. It is, therefore, very difficult to attribute unequivocally the selectivity to the bonded organic residue, except in the case of ion exchangers.

The silanol groups on the silica surface can be reacted with a variety of organic or organosilicon compounds. The reaction schemes for the preparation of chemically modified silica gels are collected in Fig.V.2.

A "monomeric" coverage of the surface with organic molecules is obtained, for instance, by esterification of the silanol groups with primary alcohols [8]. The organic moiety is actually attached by covalent bonds to the surface. Such "brushes" containing Si-O-C bonds exhibit certain advantanges for GC applications [8,9], but due to their hydrolytic instability have only limited uses for LC. Modified silica gels in which the organic moiety is bonded via nitrogen to the silicon are more stable [10,11]. Such Si-N-C bonded stationary phases can also be prepared with a large variety of organic groups by treating "chlorinated" silica gel with amines containing a second functional group (Path II). They are hydrolytically stable over the pH range of 4 to 7.5.

Chemically bonded supports with complete hydrolytic stability may be prepared by reacting silica gel with chloro or alkoxysilanes. This reaction forms a new Si-O-Si bond. An unequivocal monomeric coverage is obtained by reacting silica gels with monochloro or monoalkoxysilanes (Path IIIa). This reaction is employed in GC to eliminate the residual activity of the support for certain separation problems. Because of steric effects, the extent of reaction is quite limited and a smaller amount of carbon (i.e., functional groups) can be bonded to the silica gel than with di- or tri-functional silanes. By using dichloro or trichlorosilanes or the corresponding alkoxysilanes, the reaction can be directed along two different paths. With complete exclusion of moisture [12], i.e., by working with well-dried silica gel and very dry solvents in a dry atmosphere, the formation of polysiloxanes (bonded polymers) can be suppressed, if not completely eliminated (e.g., Path IIIb). In the case of the trichloro or trialkoxysilanes, at most two chloro or alkoxy groups react with the surface silanol groups because of steric hindrance. The unreacted groups are hydrolyzed to OH groups in the subsequent work-up, and must be resilanized with trimethylchlorosilane or hexamethyldisilazane to achieve the desired hydrophobic properties. Only after this second silanization is a negative methyl red test obtained for surface OH groups. It is also advisable to carry out this procedure when dichlorosilanes are used for the preparation of chemically bonded phases.

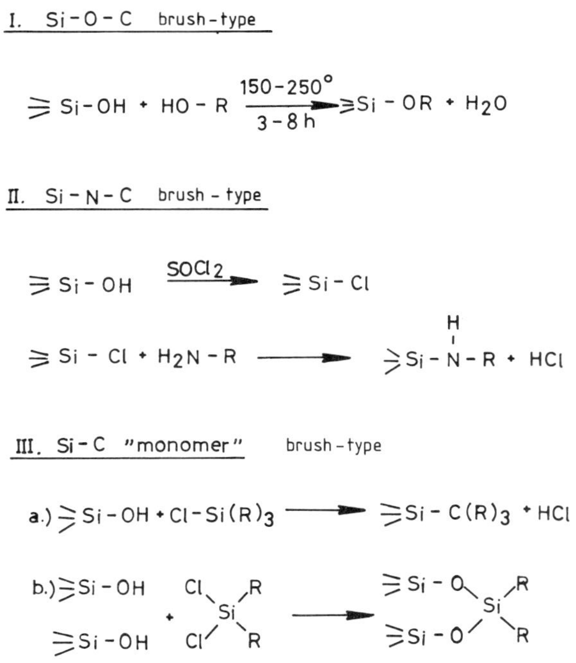

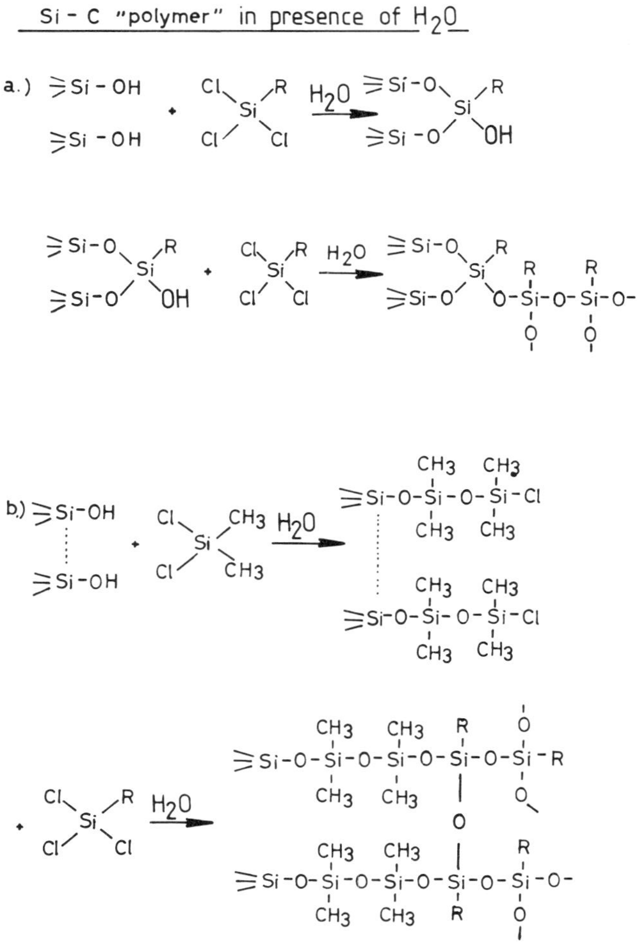

I. Si–O–C brush-type

$$\equiv Si{-}OH + HO{-}R \xrightarrow[3-8\,h]{150-250^{\circ}} \equiv Si{-}OR + H_2O$$

II. Si–N–C brush-type

$$\equiv Si{-}OH \xrightarrow{SOCl_2} \equiv Si{-}Cl$$

$$\equiv Si{-}Cl + H_2N{-}R \longrightarrow \equiv Si{-}NH{-}R + HCl$$

III. Si–C "monomer" brush-type

a.) $\equiv Si{-}OH + Cl{-}Si(R)_3 \longrightarrow \equiv Si{-}C(R)_3 + HCl$

b.) $2\,\equiv Si{-}OH + Cl_2SiR_2 \longrightarrow (\equiv Si{-}O)_2SiR_2$

Si–C "polymer" in presence of H_2O

a.) $2\,\equiv Si{-}OH + Cl_3SiR \xrightarrow{H_2O} (\equiv Si{-}O)_2Si(R)OH$

$(\equiv Si{-}O)_2Si(R)OH + Cl_3SiR \xrightarrow{H_2O} (\equiv Si{-}O)_2Si(R){-}O{-}Si(R)(O{-}){-}O{-}Si(R)(O{-}){-}O{-}$

b.) $\equiv Si{-}OH \cdots \equiv Si{-}OH + Cl_2Si(CH_3)_2 \xrightarrow{H_2O} \equiv Si{-}O{-}Si(CH_3)_2{-}O{-}Si(CH_3)_2{-}Cl \cdots \equiv Si{-}O{-}Si(CH_3)_2{-}O{-}Si(CH_3)_2{-}Cl$

$+ Cl_3SiR \xrightarrow{H_2O} \equiv Si{-}O{-}Si(CH_3)_2{-}O{-}Si(CH_3)_2{-}O{-}Si(R)(O{-}){-}O{-}Si(O{-})(O{-}){-}R$ / $\equiv Si{-}O{-}Si(CH_3)_2{-}O{-}Si(CH_3)_2{-}O{-}Si(R)(O{-}){-}O{-}Si(R)(O{-}){-}O{-}$

Fig.V.2. Reaction pathways for preparation of chemically bonded phases

On the other hand, by adding a definite amount of water (e.g., as vapor or in solution) the desired polymerization of the silane to polysiloxanes (bonded silicone oils) can be carried out [13,14](Paths IV a and b). These reactions can also be performed on PLB. Relatively thick films are obtained [15,16] by polymerization.

Whereas the chemically bonded phases with a "monomeric" coverage ("brushes") are similar to solids in their chromatographic properties (e.g., the rate of mass transfer), those with bonded polysiloxanes rather resemble partition systems in which the solid is coated with a liquid phase that is insoluble in the eluent. Since the polymerization in part produces highly cross-linked polysiloxanes, the diffusion in the polymers is slow and restricted, and the mass transfer is slow, i.e., the H values rise steeply with increasing flow rate [16]. In practice, however, it is not possible to clearly differentiate between "monomerically" and "polymerically" bonded phases. Which of these two types has advantages depends on the given system [17]. However, chemically bonded stationary phases possess certain advantages over conventional particles coated with liquid phase:

1) The eluent need not be saturated with the liquid phase.

2) No precolumn is necessary.

3) In sample isolation from the eluent there is no contamination from the stationary phase, a condition that usually hampers preparative application of partition systems.

4) The stationary phase cannot be stripped off because it is covalently bonded to the solid. The problem of mechanical erosion at higher flow rates is eliminated.

5) Various programming techniques, including gradient elution, can be employed, which is not the case for conventionally coated supports.

6) Column equilibrium is not so readily disturbed by external effects (temperature, eluent moisture, etc.).

The last two points also demonstrate their advantages with respect to ordinary polar stationary phases such as silica gel and alumina, for which regeneration following gradient elution often takes longer than the gradient run itself because of the very slow attainment of equilibrium with the water that is inevitable present in the eluents.

If the organic moieties containing no functional groups, i.e., alkylsilanes, are used (ordinarily with chain lengths of 4 to 18 C-atoms), stationary phases for reversed phase chromatography are obtained. In contrast to conventional adsorption chromatography (cf. Chapter VI), in reversed phase systems the stationary phase is nonpolar (hydrophobic) and the mobile phase is polar (hydrophilic). The reversal of the phase properties also reverses the sample elution order, so that nonpolar compounds are more strongly retarded than the polar. Such systems are preferable for samples that dissolve only in relatively polar solvents, from which, however, they are no longer retained on polar adsorbents. In addition, nonpolar compounds (such as fats) that are not retarded on polar adsorbents by nonpolar eluents can be separated on reversed phase systems. True reversed phase systems are hydrophobic, are not wetted by water, and should have no silanol groups that are accessible to sample molecules. Free, unreacted, and unshielded silanol groups can be easily identified by the adsorption of methyl red from a benzene solution [23]: if the silica gel retains a red violet color after being shaken with a methyl red solution and washed with pure benzene, free silanol groups are present. In addition, the retention times of polar compounds using nonpolar eluents should be very small on "true" reversed phases [18].

Other chemically bonded stationary phases can be prepared by utilizing chloro or alkoxysilanes with the appropriate functional groups [24]. Stepwise preparation of derivatives starting with reversed phase systems is also possible [12]. These include, among others, cation exchangers (with alkyl or arylsulfonic acid groups) and anion exchangers (with amino, dialkylamino, or quarternary ammonium groups). The selectivity of chemically bonded stationary phases containing other functional groups is difficult to determine and should be established with the greatest of care because the number of unreacted silanol groups frequently exerts a greater effect on selectivity than those functional groups which were introduced with considerable effort.

In the meanwhile a great variety of chemically bonded phases has become commercially available as such or only as packed columns, in large (30 - 50 μm) and small particle sizes (around 5 or 10 μm). Besides alkyl groups of different chain lengths (commonly octyl or octadecyl groups), aryl, alkylated amino, nitrile, nitro, diol groups, etc., have also been bonded to the silica surface.

The great versatility of chemically bonded phases has recently prompted the appearance of numerous review articles and papers [17-20, 25-33]. The applications of these phases are discussed in Chapter VI.

C. Ion Exchangers

The classical porous ion exchangers have only limited utility for HPLC because of their compressibility. Recently, quite pressure-stable ion exchangers with particle sizes < 20 μm have been specially developed to improve amino acid analysis. These have enabled separations to be performed at higher pressures (up to about 200 atm). The capacity of these exchangers is about 3 - 5 meq/g, i.e., approximately 100 times greater than that of exchangers coated on PLB. The volume of ordinary exchangers varies with the pH, ionic concentration, and temperature of the eluent. They can only be packed in a pre-swollen state into a column.

The first ion exchangers especially developed for HPLC were prepared by polymerizing a polystyrene film on solid glass beads, followed by the introduction of functional groups [34]. The capacity of these so-called "pellicular" ion exchangers is "adequate". In nonaqueous solvents the polymer layer may swell or dissolve. The same precautionary measures should be taken when using ion exchangers coated on Zipax®. The methacrylate polymer is soluble in organic solvents.

Other cation and anion exchangers based on PLB are commercially available in particle sizes normally used for PLB (between 30 and 50 μm). Almost all of these ion exchange particles are spherical and possess good chromatographic properties because of the thin diffusion layer. Their greatest disadvantage lies in their low capacity, which amounts to 10 - 50 μeq/g. Since a change in the ionic strength or pH value does not alter their volume, they are well suited for use in columns. The permeability of such columns scarcely changes during operation. Exchangers are prepared by chemical substitution of suitable silane derivatives, either prior to or following dispersion on the support particles. Their pH stability is adequate.

However, in alkaline media (pH > 8) the attack by hydroxide ions on the silica support must be taken into account. It should be pointed out that the strongly basic anion exchangers on silica gel, if present in the hydroxide form, catalyze their own dissolution in water.

Recently, ion exchangers have been described and become commercially available in which alkyl or aryl groups are covalently bonded to porous silica gel (in the manner of "brushes") and onto which ion exchange groups have been introduced [35-39]. Starting with small particles (5 - 10 μm) of silica gel, ion exchangers with good chromatographic properties and adequate capacity (200 - 1000 μeq/g) are obtained. Since the capacity is a function of the specific surface area of the starting silica gel, the capacity of such exchangers can be adapted to practically any separation problem.

D. Stationary Phases for Exclusion Chromatography

In exclusion chromatography the separation is based on the varying degrees of accessibility to the inner pore structure of the support (cf. Chapter IX). Originally, polymers (= gels) swollen with eluent were used for this type of chromatography. The materials utilized for aqueous systems included cross-linked dextrans (Sephadex®), polyacrylamides (Bio-Gel P®), or agarose (Bio-Gel A®). For the separation of hydrophobic polymers with organic eluents polystyrenes of various degrees of cross-linking (Styragel®, Poragel®, and Bio-Beads S®) have been employed. Cross-linked polyvinyl acetate gels, such as Merckogel Type OR® are also useful. These hydrophobic gels exhibit good separation characteristics only if the eluent wets the support surface and swells the polymer.

The gels (swollen polymers) are not particularly pressure-stable, and therefore separations can only be carried out at very low eluent velocities (small pressure drop). Moreover, the swelling and shrinking of the packing inside the column may lead to a deterioration of the permeability and/or separation efficiency. Consequently, the "classical" gels are hardly suitable for exclusion chromatography under HPLC conditions.

However, if fine sieve fractions ($d_p < 15$ µm) of these gels are used, exclusion chromatography can be carried out at higher separation velocities. Polystyrenes of this size are commercially available under the name of Styragel® (Waters Associates, Milford, Mass.). A few other gels, such as Merckogel Type OR® or the hydrophilic Merckogel PGM® polyethylene glycol methacrylate [40] are stable up to moderate pressures (~ 60 atm). Spheron®, also a polyethylene glycol acrylate, is suitable for aqueous gel chromatography.

No problems with pressure stability arise with porous glasses and silica gels. Both involve nearly pure SiO_2; only the pores are produced in different ways. Silica gel is made either from water glass or by polymerization of tetraethyl silicate. The existing pore diameters can be increased by subsequent hydrothermal treatment [4]. Porous glasses are obtained from demixed borosilicate glasses by removing the separated boric acid with steam. Both supports have acidic groups on the surface (silanol or Lewis-acid groups) whose interaction with certain samples may be significant (e.g., the tertiary structure of proteins may be destroyed).

Silica gels (such as Lichrosorb®, Lichrospher®, Porasil®, Spherosil®) and porous glasses (such as CPG®-10 or "Bio-Glass"®) are commercially available with pore diameters from 60 - 2500 Å. Lichrosorb® is available with pore sizes up to 25 000 Å. Both polar and nonpolar eluents may be used with silica gel and porous glass. With nonpolar eluents care must be taken that the samples do not adsorb on the support surface. Chemically modified silica gels [24,27,41-44], with glycol ether or other end groups, are used for the exclusion chromatography of biological substances. These phases are wetted by water, but a reversed phase interaction may occur in addition to the exclusion.

References Chapter V

General description of commercially available supports and columns for HPLC and gel chromatography:

Majors, R.E.: Recent Advances in High Performance Liquid Chromatography Packings and Columns, J. Chromatogr. Sci. *15*, 334-351 (1977)

1. Eisenbeis, F.: Poster Presentation 3. Int. Symposium on Liquid Chromatography, Salzburg 1977
2. Karger, B.L., Engelhardt, H., Conroe, K., Halâsz, I., in: Stock, R. (Ed.): Gas Chromatography 1970. London: Institute of Petroleum 1971
3. Unger, K.: Angew. Chemie *84*, 3311 (1972)
4. Iler, K.: The Colloid Chemistry of Silica and Silicates. Ithaka: University Press 1965
5. Courtesy Dr. F. Eisenbeis, Merck AG, Darmstadt
6. Snyder, L.R.: Principles of Adsorption Chromatography. New York, N.Y.: Dekker 1968
7. Uihlein, M.: Dissertation, Saarbrücken 1971
8. Halâsz, I., Sebestian, I.: Angew. Chemie *81*, 464 (1969)
9. Halâsz, I., Sebestian, I.: J. Chromatogr. Sci. *12*, 161 (1974)
10. Brust, O.E., Sebestian, I., Halâsz, I., in: Perry, S.G.: Gas Chromatography 1972. Ec. Applied Science. Barking, Essex 1973
11. Brust, O.E., Halâsz, I.: J. Chromatogr. *83*, 15 (1973)
12. Sebestian, I., Halâsz, I.: Chromatographia *7*, 371 (1974)
13. Aue, W.A., Hastings, C.R.: J. Chromatogr. *42*, 319 (1969)

14. Hastings, C.R., Aue, W.A., Augl, J.M.: J. Chromatogr. *42*, 487 (1970)

15. Kirkland, J.J., DeStefano, J.J.: J. Chromatogr. Sci. *8*, 309 (1970)

16. Kirkland, J.J.: J. Chromatogr. Sci. *9*, 206 (1971)

17. Sebestian, I., Halâsz, I., in J.C. Giddings et al.: Advances in Chromatography, Vol. 14. New York: Dekker 1976, p. 75

18. Karch, K., Sebestian, I., Halâsz, I.: J. Chromatogr. *122*, 3 (1976)

19. Grushka, E., ed.: Bonded Stationary Phases in Chromatography. Ann Arbor, Michigan, U.S.A.: Ann Arbor Science Publ. 1974

20. Řehák, V., Smolkova, E.: Chromatographia *9*, 219 (1976)

21. Hesse, G.: Z. Anal. Chem. *211*, 5 (1965)

22. Hesse, G.: Chromatographisches Praktikum. Frankfurt/Main: Akadem. Verlagsgesellschaft 1968

23. Kolthoff, I.M., Shapiro, I.: J. Am. Chem. Soc. *72*, 776 (1950)

24. Engelhardt, H., Mathes, D.: J. Chromatogr. *142*, 311 (1977)

25. Kirkland, J.J.: Chromatographia *8*, 661 (1975)

26. Boksanyi, L., Liardon, O., Kovats, E. sz.: Adv. Coll. Interfac. Sci. *6* 95 (1976)

27. Chang, S.H., Gooding, K.M., Regnier, F.E.: J. Chromatogr. *120*, 321 (1976)

28. Unger, K., Becker, N., Roumeliotis, P.: J. Chromatogr. *125*, 115 (1976)

29. Horvâth, C., Melander, W., Molnâr, J.: J. Chromatogr. *125*, 129 (1976)

30. Karger, B.L., Gant, J., Hartkopf, A., Weiner, P.H.: J. Chromatogr. *128*, 65 (1976)

31. Collin, H., Guichon, G.: J. Chromatogr. *141*, 289 (1977)

32. Molnâr, J., Horvâth, C.: J. Chromatogr. *142*, 623 (1977)

33. Scott, R.P.W., Kucera, P.: J. Chromatogr. *142*, 213 (1977)

34. Horvâth, C., Preis, B.A., Lipsky, S.R.: Anal. Chem. *39*, 1422 (1967)

35. Unger, K., Nyamah, D.: Chromatographia *7*, 63 (1974)

36. Weigand, N.: Dissertation, Saarbrücken 1974

37. Weigand, N., Sebestian, I., Halâsz, I.: J. Chromatogr. *102*, 325 (1974)

38. Saunders, D.H., Barford, R.A., Magidman, P., Olszewski, L.T., Rothbart, H.L.: Anal. Chem. *46*, 834 (1974)

39. Asmus, P.A., Low, C.-E., Novotny, M.: J. Chromatogr. *119*, 25 (1976)

40. Heitz, H.: Angew. Chemie *82*, 675 (1970)

41. Chang, S.H., Gooding, K.M., Regnier, F.E.: J. Chromatogr. *125*, 103 (1976)

42. Wu, A.C.M., Bough, W.A., Conrad, E.C., Alden jr., K.E.: J. Chromatogr. *128*, 87 (1976)

43. Regnier, F.E., Noel, R.: J. Chromatogr. Sci. *14*, 316 (1976)

44. Persiani, C., Cukor, P., French, K.: J. Chromatogr. Sci. *14*, 417 (1976)

Chapter VI

Adsorption Chromatography

I. Polar Stationary Phases

A. General

By far the most LC separations are based on adsorption effects. Such separations are governed by the "adsorption milieu": the interaction of adsorbent, solute, and eluent. For oxidic sorbents the sorption is based on specific interactions between the polar support surface and the polar groups of the adsorbed molecule. These include dipole-dipole interactions between permanent and induced dipoles, hydrogen bonding, as well as charge transfer or π-complex formation. Chemisorption, which occurs occasionally, is generally undesirable because it can lead to extremely long retention times, irreversible sorption, or even sample decomposition.

A *linear sorption isotherm* is essential for reproducible chromatographic work. Only then is the retention time independent of the sample size. Unfortunately, in all practical cases the sorption isotherms are more or less curved. These are primarily Langmuir-type isotherms in which the relative amount adsorbed decreases with increasing sample concentration, i.e., the retention time decreases (Fig.VI.1). Furthermore, the zone of highest concentration is shifted more and more to the front of the sample band (region of saturation concentration) and gives rise to asymmetric elution peaks (tailing). In adsorption chromatography such asymmetric distributions can be attributed almost exclusively to nonlinear isotherms.

If separations are carried out in the nonlinear region of the isotherm, the retention times depend on the sample size, which adversely affects qualitative identification of the components. Tailing may cause the resolution of two adjacent zones to be incomplete, which may hamper or prevent isolation of the pure components.

Experience has shown the linear region of the isotherm to be almost identical for many adsorbents. It can be assumed that for sam-

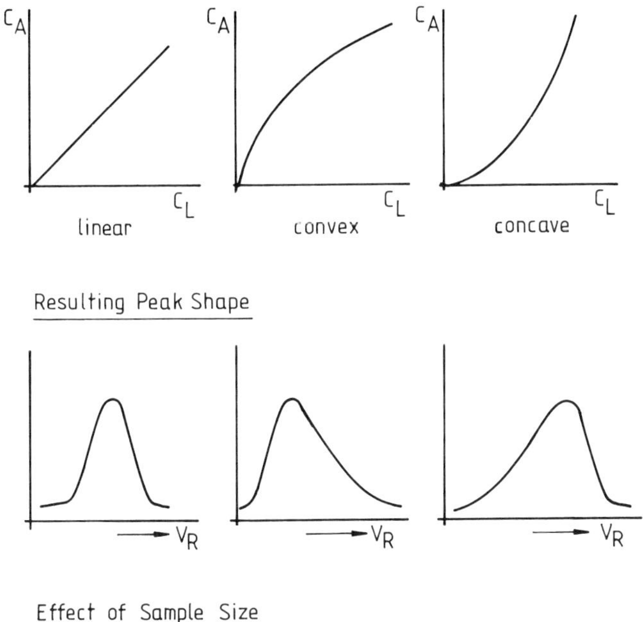

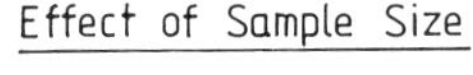

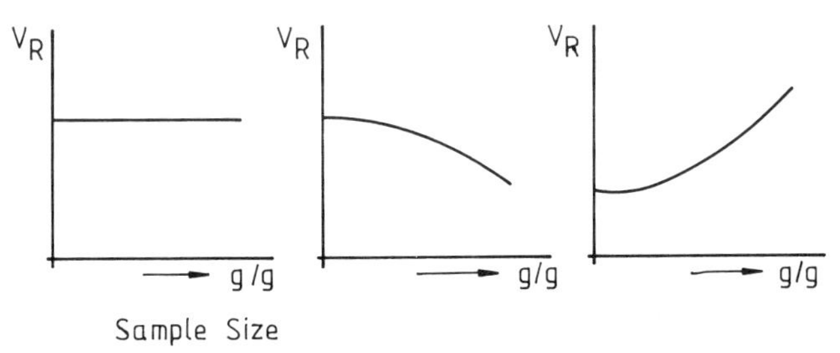

Fig.VI.1. Effect of the sorption isotherm on peak shape and retention

ple sizes of less than 10^{-4} g sample/g of adsorbent the retention times are independent of sample size and that one is therefore operating in the linear region of the isotherm [1]. This region is also called the linear capacity of an adsorbent. The linear capacity decreases as the adsorbent surface becomes more inhomogeneous. By coating the surface with a polar compound such as water or simply adding it to the eluent, the linear region of the isotherm is increased.

In addition to deactivation with water or other polar substances, surface modification by chemical reaction can also be used to extend the linear region. Thus, for good reversed phases the load capacity is at least an order of magnitude greater than that of the starting silica support. This type of surface homogenization has the advantage that the chemically bonded compound is not stripped off when the eluent is changed.

In critical cases the linear capacity of the adsorbent should be determined experimentally. To do this, the k' or H values are plotted against the logarithm of sample size per gram of adsorbent. Fig. VI.2 presents such an experimental curve for silica gel. As is clearly evident, both the H and k' values change significantly at a certain sample size (10^{-4} g/g). According to Snyder [1], a 10% change in the k' value can be tolerated (definition of the linear capacity). This value from classical column chromatography appears to be somewhat too large for HPLC. At any rate, the established maximum sample size should not be exceeded if identification is based on retention volumes. The increase in band width may be attributed to tailing.

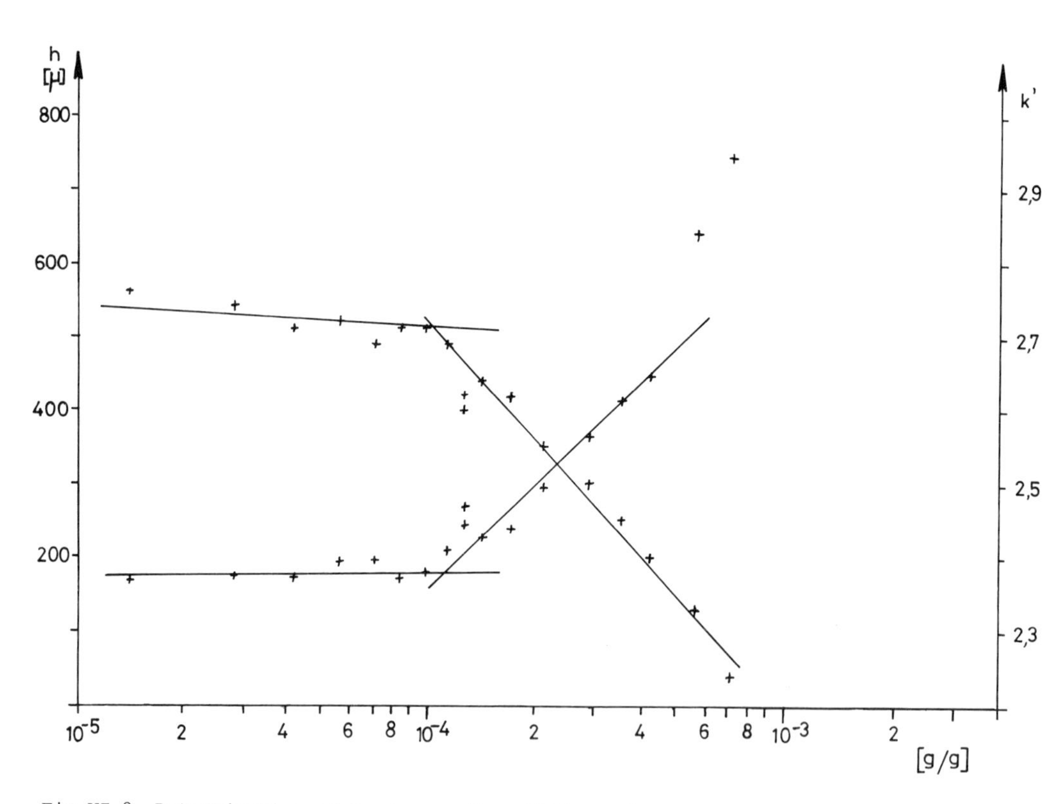

Fig.VI.2. Determination of load capacity. Silica gel, Merckogel Si 100 (d_p ~ 10 μm); eluent: n-heptane (40 ppm H_2O); sample: nitrobenzene

In HPLC the sample volume injected does not play as important a role as it does in GC. Sample volumes of up to 50 μl have little, if any, effect on column efficiency under ordinary conditions, but if the mobile phase volume is small and the k' values are low (< 2) some deterioration may be observed. For components with higher k' values and in gradient elution the sample volume is not a factor. However, to avoid difficulties the standard rule should be to dissolve the sample in the smallest volume that can be reasonably handled.

The tailing attributable to a nonlinear isotherm can be distinguished from that due to instrumental factors, such as excessive dead volume. Tailing caused by instrumental deficiencies is particularly pronounced for the unretained substance as well as those with low k' values (< 3); those with high k' values generally exhibit no tailing (a linear sorption isotherm is assumed in this case). On the other hand, if the tailing is due to a nonlinear isotherm, the unretained peak remains symmetrical. For such an isotherm, peak asymmetry increases with rising k' values, a contributing factor in this case being the increased amount of sample necessary to elicit a measurable detector signal.

A concave sorption isotherm, as depicted in Fig.VI.1, is very rare. In GC it may occur as a result of adsorption on a liquid interface. A chromatographic peak characterized by a slow rise and a steep descent is designated as "leading". Such peaks may also be obtained when the sample is sparingly soluble in the eluent. Isotherms that are completely linear over an extended concentration range are rare. For any new system, it is therefore recommended that the independence of the retention times on sample size be established experimentally. Only then it is possible to identify the components of a mixture on the basis of their retention times.

B. Stationary Phases

As discussed in Chapter V, primarily silica and, to a lesser extent, alumina are employed in HPLC as polar stationary phases, either directly as a porous support or in the form of PLB. In addition to these polar systems, chemically modified phases - predominantly reversed phases with a nonpolar surface - are being used increasingly today. Their area of application corresponds to that of activated charcoal in classical column chromatography. Because of their poor mechanical properties (and poor reproducibility in preparation), activated charcoals are unsuitable for HPLC without modification. The polar functional groups (phenolic, carboxylic, etc.) on its surface enable charcoal to act as a polar adsorbent when used in conjunction with non-polar eluents. A more detailed discussion of reversed phases follows in part II of this chapter.

The silica used for adsorption chromatography should generally have a high specific surface area, preferably > 200 m^2/g, and an average pore diameter > 60 Å, in order to achieve high k' values. As is illustrated in Fig.V.1, there is a good correlation between the surface area of the silica and the absolute retentions. At times, however, lower k' values may be desired, and these are usually attained by eluting with a more polar solvent (cf. Section D), although low sample solubility may restrict this approach. In such case low-surface silicas or PLB can be used to advantage.

It bears repetition that the active adsorbents are more or less amorphous materials whose properties often vary from batch to batch. Hence, it is no wonder that the relative and absolute retentions vary when silicas from different suppliers are compared. For example, the relative and absolute retentions of m- and p-nitroaniline were determined with three different brands of silica under identical chromatographic conditions. For the first two brands the relative retentions α were the same (1.36), but the k' values were about 1 and 2. The second and third exhibited similar k' values (~ 2), but the α of the third was only 1.08 [2]. Inversion of the elution order of some components upon changing brands of silica is not uncommon.

In general, similar elution orders are observed on silica and alumina. However, the latter is better suited for the separation of condensed aromatic hydrocarbons. This is illustrated dramatically in the separation of anthracene and chrysene on silica and on alumina. By varying the water content of the eluent (heptane), k' values for anthracene ranging from 0.9 to 3.7 were obtained on both adsorbents. But whereas the α value for this pair remained constant at 1.9 on silica, it increased strikingly from 2.3 to 8.6 on alumina. This indicates that the retention mechanism of these hydrocarbons remains the same on silica but certainly changes on alumina.

Silica coated with silver nitrate selectively separates saturated from unsaturated organic compounds. This well-known system from classical column chromatography has also been occasionally used in HPLC [3]. Other compounds that interact with olefins, such as trinitrofluorenone, also yield very selective columns [4].

Chemically modified stationary phases with appropriate functional groups exhibit a selectivity analogous to that of physically coated phases [5]. These phases have all of the advantages of chemically bonded stationary phases as compared to physically coated ones.

C. Effect of Water on Separations

The oxide adsorbents such as alumina and silica gel are known to be good drying agents for nonpolar organic solvents. The absorbed water exerts a considerable effect on the chromatographic properties. This fact was recognized early and numerous attempts were made to standardize the water content of these adsorbents [7,8,9] and to relate this to the equilibrium water content of eluents [9,10]. These procedures, in which definite amounts of water are added to an adsorbent having a certain initial activity, have only limited applicability for HPLC. For the considerably greater amount of eluent flow per gram of adsorbent, as compared to that in classical column chromatography, the water content of the eluent governs the adsorbent activity more strongly than the previously added water. Rapid equilibration is achieved between the adsorbed water and that dissolved in the eluent. If the columns are packed by one of the suspension techniques describend earlier, the activity can only be adjusted by way of the eluent.

The effect of the ubiquitous water on the separation is illustrated in Fig.VI.3 [11]. In this example, an optimum separation of oligophenyls was achieved by using a mixture of one part water-saturated heptane with two parts dry heptane as eluent (b). With an excessively high water content (c) separation was no longer possible.

The opposite effect may occur, where the water content of the eluent enables a separation to be attained. Fig.VI.4 a shows the separation of steroids on silica gel using chloroform dried over molecular sieves. The separation is incomplete and the peaks exhibit strong tailing. In contrast, when water-saturated chloroform (~ 0.1% water) is employed, all three substances are well resolved and eluted as sharp peaks in less time (Fig.VI.4 b). In this case a partition system is formed with water as the stationary liquid phase (cf. Chapter VII.B.3).

Whether desirable or not, each eluent contains a certain amount of water. Even the commercial "pure solvents" have significant but

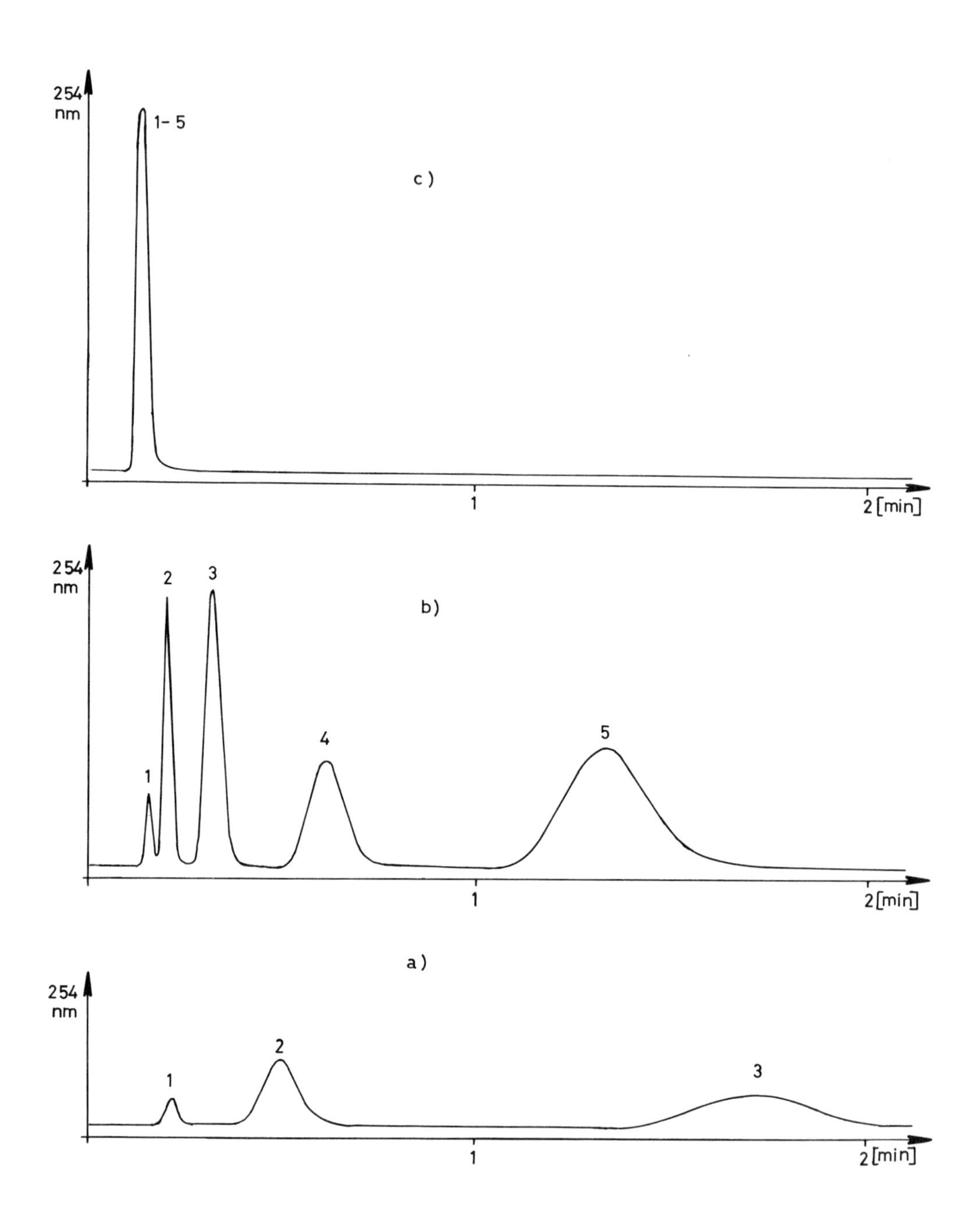

Fig.VI.3. Effect of moisture content of the eluent on separation [11].
Stationary phase: Perisorb A (PLB); $d_p \sim 30$ µm; eluent: a) n-heptane, dried over molecular sieve, b) 100 ml water-saturated heptane + 200 ml heptane dried over molecular sieve, c) water-saturated heptane; column: 50 cm, 2 mm i.d.; u = 6.2 cm/sec; Δp = 56 atm; samples: 1 = benzene; 2 - 5 = m-oligophenyls

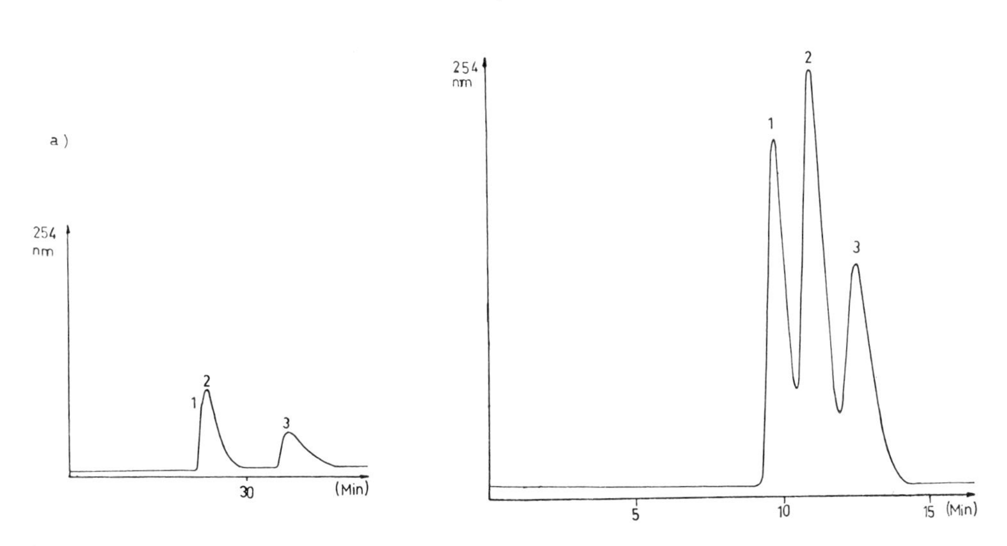

Fig.VI.4. Separation of steroids on silica gel [11]. Merckogel Si 60; eluent: chloroform + 0.5% alcohol as stabilizer; a) dried over molecular sieve, b) water-saturated (~ 0.1% H_2O). Column: 200 cm, 2 mm i.d.; F = 0.75 ml/min; Δp = 185 atm; samples: 1 = dexamethane; 2 = fluorhydrocortisone; 3 = hydrocortisone

variable amounts of moisture. To maintain an eluent with a definite water concentration is almost impossible because the dry or wet container surface alone may alter the concentration appreciably. Furthermore, changes in the water content on opening the flask or pouring the eluent into another container are virtually inevitable.

Other polar impurities or admixtures, even in trace amounts, exert a similar effect. It should be borne in mind that small amounts (0.2 - 2%) of ethanol are added to halocarbons as stabilizer. Phenols serve a similar function in ethers such as dioxane and tetrahydrofuran. Because these stabilizers may have considerable influence on separations, adsorption chromatography has sometimes been characterized as being poorly reproducible. Hence, only purified and dried solvents of specified quality should be used as eluents; purification by passing them over activated adsorbents [9,12] should be made a standard procedure in every chromatographic laboratory.

The effect of changes in the moisture level of eluents on sample retentions is most pronounced for the least polar solvents, such as the hydrocarbons, in which water has a very limited solubility ($\leq$ 100 ppm). Conversely, the greater the solubility of water, the smaller is this effect. For example, the water content of methylene chloride (solubility *ca.* 0.2%) may vary by several ppm without appreciably changing the k' values.

The reproducibility of k' values can be increased by monitoring and controlling the water content of the mobile phase and continuously recycling a sufficiently large volume in a closed system to maintain equilibrium. Ordinarily the water content may be determined by a Karl Fischer titration, but for the low concentrations in aliphatic hydrocarbons this method is fraught with difficulties. To obtain even modest accuracy (± 5 ppm at H_2O levels < ppm; DIN 51777, G.F.R.) relatively large eluent volumes are required (⩾ 200 ml). Significantly, at these levels the k' values show a sensitive dependence on fluctuations in the water content that cannot be detected by the Karl Fischer titration [13].

One way to circumvent these problems with the water determination was proposed by Snyder [14]. He recommended the use of eluents with the same percentage of water saturation, say 50%. These partially water-saturated eluents can be prepared by blending corresponding amounts of "dry" and "completely water-saturated" mobile phases. In practice, however, the defined status "dry eluent" involves some of the problems already discussed. On the other hand, it can be surprisingly difficult to completely saturate solvents having a low water solubility (especially hydrocarbons and their mixtures) by simply shaking or stirring them with water [15].

After conditioning a column with such an eluent, an equilibrium is established between the water in the eluent and that on the adsorbent, thereby yielding a constant and reproducible water coating. Since this equilibrium coating remains unchanged when other eluents with the same degree of water saturation are substituted, the usually lengthy re-equilibration times associated with solvent changeover can be avoided. Similarly, the initial equilibration times with nonpolar eluents may be shortened by first conditioning the column to the desired water-saturation with a mobile phase having an appreciable solubility for water, followed by the nonpolar eluent with the same relative saturation. For optimum surface coverage [1], assuming monolayer water adsorption, 25% water-saturated eluents are recommended for alumina and 50% saturated for silica [15].

Another convenient way is to condition eluent and column together by recycling the eluent through an apparatus for buffering the moisture in the entire chromatographic system, called a moisture control system (MCS). Such MCS can be readily assembled and installed before the pump [13].

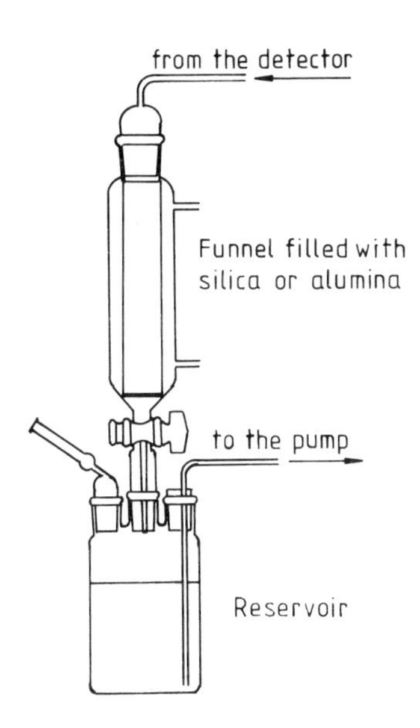

Fig.VI.5. Sketch of the moisture control system

An apparatus assembled from commercially available glass components is shown in Fig.VI.5. The MCS consists of a 500 - 1000 ml Woulff flask equipped with a thermostated dropping funnel (100 - 200 ml) that has a glass frit or perforated plate (for use with filter paper) sealed in at the bottom. The dropping funnel is filled with approximately 100 g of coarse alumina or 50 g of silica gel that was either dried (activated) or coated with known amounts of water (between 3 - 20% w/w). The amount of silica or alumina must be increased for conditioning eluents having a higher solubility for water, such as dichloromethane. As a rule of thumb, the amount of water dissolved in the recycled eluent should not exceed 1% of that held on the adsorbent in the MCS funnel.

For convenience, the quantities of water may be used that correspond the amounts required to prepare the Brockmann activity grades [7]. However, to optimize a separation, the adsorbent in the funnel may be precoated with any desired amount of water (cf. Fig.VI.7).

Depending on the initial water content, the eluent is either dried or moistened to an equilibrium value. The amount of water on the silica or alumina in the funnel as well as the temperature determine the absolute and relative water content of the eluent, and consequently the absolute and relative sample retentions obtained with the column. The time required to attain equilibrium depends, of course, on the volume flow rate of the eluent and the solubility of water therein. Longer times are always required to "dry" a column (to increase k' values) than to run "wet". As a guideline for n-heptane, approximately 18 hours are required to run "wet", whereas 48 hours are needed to attain extreme dryness (for a flow rate of about 4 m/min). For methylene chloride the duration required is substantially shorter (6 - 14 hrs). The injected sample components are subsequently removed by the MCS and do not interfere in other separations. Several hundred separations may be carried out without difficulty with a single MCS charge (500 - 1000 ml).

As in the previous method, the equilibration times with hydrocarbon eluents can be shortened by first conditioning the column with an eluent having an appreciable solubility for water. After the column has attained equilibrium, a second MCS already pre-equilibrated with the desired eluent is connected in place of the first. Both MCS should hold adsorbent containing the same amount of water.

An increase in the temperature of the MCS funnel raises the water concentration of the recycled eluent. Consequently, the k' values of the sample components decrease, if the temperature of the chromatographic column is kept constant.

Fig.VI.6 shows a separation of condensed aromatic hydrocarbons in which the water content of the eluent (n-heptane) was controlled by using the MCS. The water concentrations determined by Karl Fischer titration are included for comparison. If all variables (water content of the silica in the MCS, temperature of MCS and chromatographic column) are kept constant, the reproducibility of the k' values is $\pm$ 2.5% (k' < 4).

Table VI.1 presents the ranges of water concentrations in various eluents [10,13,16]. For the sake of convenience the coverages of the alumina in the MCS were chosen to correspond to the Brockmann [7] activity grades. In addition, the system can be characterized further by means of the k' values of several standard substances.

In some cases the recycling method cannot be employed on account of the system or the apparatus. In such cases the separation system can only be standardized by means of chromatographic data. To do this, a standard mixture of samples with k' values between 1 and 10 is prepared. The k' values of this standard mixture are determined with each new batch of eluent as soon as the column has been equilibrated. Higher k' values would indicate that this new batch of eluent is drier than

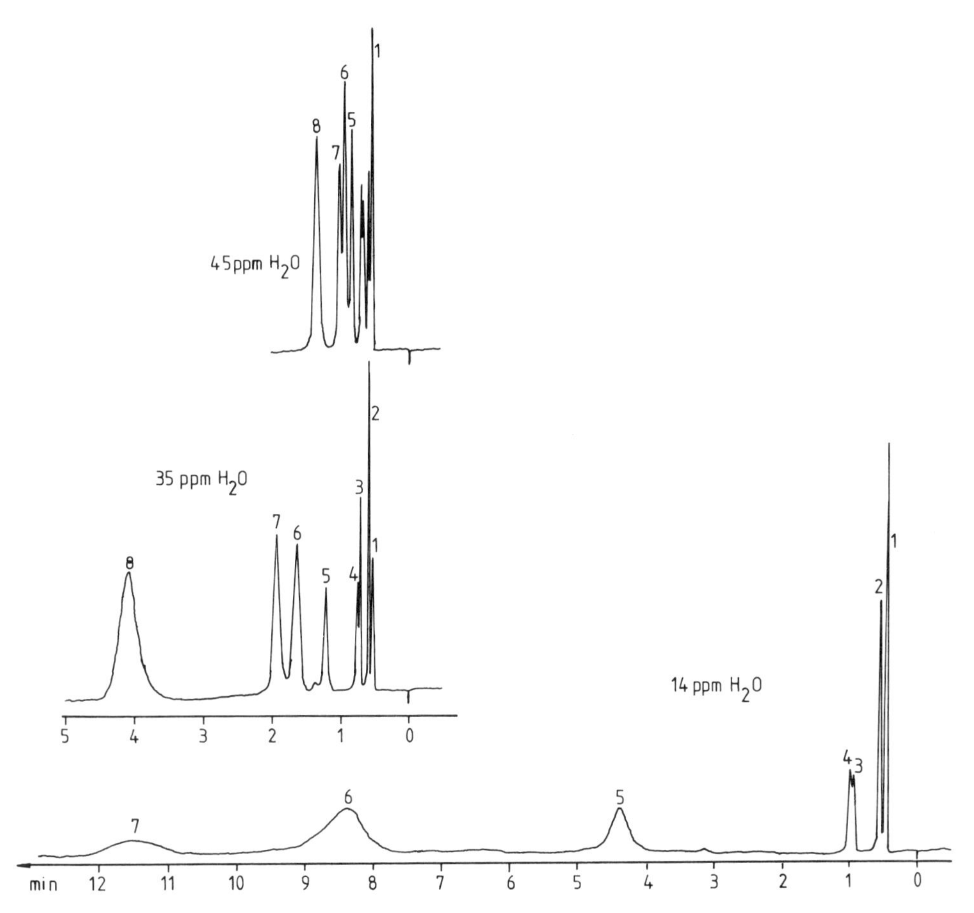

Fig.VI.6. Separation of aromatics with various water concentrations in the eluent, adjusted via the MCS. MCS: alumina Woelm neutral for CC, coated with 4.5, 6, and 9% (w/w) water; temperature: 25°C; eluent reservoir 500 ml. Analytical column: alumina Woelm neutral; d_p ~ 5 μm; t = 25°C; eluent: n-heptane with 14, 35, and 45 ppm water. Δp = 100 bar; F = 2.8 ml/min; u = 4.2 mm/sec. Samples: 1 = Inert; 2 = benzene; 3 = naphthalene; 4 = biphenyl; 5 = anthracene; 6 = pyrene; 7 = fluoranthene; 8 = 1,2 benzanthracene

the original. The k' values may then be reduced to the desired values by the addition of water-saturated eluent. If the k' values are lower than the original, an accordingly more strongly dried eluent must be used. Particular care is advisable for separations performed in systems involving adsorbents with a high specific surface area and non-polar eluents with relatively low water contents.

As is evident from Table VI.1, each solvent has a different equilibrium water content for a specific activity grade. Consequently, in gradient elution the situation is particularly complex because the amount of adsorbed water is constantly changing. Depending on the

Table VI.1. Water concentrations of eluents in equilibrium with alumina coated with H_2O

	Extent of coating of Al_2O_3				
	0	3	6	10	15 %
n-Heptane	< 5	15	35	55	60
Carbon tetrachloride	< 5	40	120	150	160
Benzene	< 5	100	340	400	460
Di-isopropyl ether	< 10	250	1050	2400	3200
Chloroform	< 10	140	700	900	1200
Methylene chloride	< 10	700	1200	1500	1800

water content of the polar solvent, water is furnished or removed by the adsorbent. On returning to the original solvent, some time is required to restore the initial conditions with respect to the adsorbed quantities of water.

Because of the problems encountered with these water systems, the use of primary alcohols (methanol, ethanol, isopropanol) as modifiers of the adsorption strength (called moderators) has been recommended [14,17,18,19]. Usually, between 0.01 and 0.5 v/v% alcohol is added to the eluent. As an example, the k' values for the benzyl alcohols on a silica column are in the same range when eluted with dichloromethane containing either 0.1% water (50% water-saturated) or 0.15% methanol or 0.3% isopropanol [19]. The preparation and preservation of these alcohol-eluent mixtures is accompanied by problems similar to those discussed with water-moderated eluents. No information could be found in literature on how changes in the water content of the original eluent additionally influence the k' values measured in alcohol-moderated systems. Some additional problems such as distorted or asymmetric peaks, less efficient columns etc., have been reported for alcohol-moderated systems [14,19].

In addition to water or alcohols, other more or less polar organic compounds may be added to nonpolar eluents to adjust the k' values of given samples. As the polarity of the added moderator decreases, its amount in the eluent must be increased. If the moderator concentration is 1% or greater it is better to speak of eluent mixtures

(Section D.3) than of moderated systems. Most of the problems involving "moderated eluents" stem from the difficulties inherent in determining and reproducing the (very) small quantities of polar substances that are usually added to the eluent.

The effect of solvent composition on selectivity has been well established [1] and will be discussed later. It should be pointed out here that even the small moderator concentrations affect not only absolute but also relative retentions and that selectivity is not independent of the type of modifier used [20,21,22].

The rapid attainment of equilibrium between the concentration of a polar compound dissolved in an eluent and adsorbed on an active adsorbent can be used to load an active support with variable amounts of stationary phase. Eventually, a column containing an active adsorbent can be transformed into a "partition column". In some cases this is the only way to prepare a partition-chromatographic system. In fact, this is a practical way of coating an active solid with any desired polar liquid phase. The polar components are adsorbed preferentially from ternary mixtures and form the stationary liquid phase (cf. Chapter VII).

In summary, the addition of small amounts of water or other polar moderators to an adsorbent or eluent reduces the retention volumes to the extent that nonpolar compounds are no longer retained. By "activating" the absorbent on the column, preferably by utilizing drier or purified eluents, the substances that were not retained in the "wet" system will be retarded and separated. In most cases the elution sequences remain unchanged.

D. Effect of Eluent on Separation

The choice of the proper solvent frequently affects the success of a separation more than the selection of the stationary phase. Depending on the properties of the eluent, on a given adsorbent a sample may be excessively retained or not at all, or its retention time may fall into the desired range. Beyond this, however, the eluent must also meet the following requirements.

It must be possible to detect the sample in the column effluent. When a UV detector is used, for example, the eluent should not absorb at the detector wavelength.

Table VI.2. Eluotropic Series. Properties of important solvents for adsorption chromatography [a]

	Solvent strength [1] ε_0	Dielectric constant	Viscosity η [c.P.] (20°C)	Refractive index (20°C)	Lowest useable wavelength [nm]
n-Pentane	0.00	1.84	0.235	1.358	200
n-Hexane	0.01	1.88	0.33	1.375	200
n-Heptane	0.01	1.92	0.42	1.388	200
Isooctane	0.01	1.94	0.50	1.391	200
Cyclohexane	0.04	2.02	0.98	1.426	210
Carbon tetrachloride	0.18	2.24	0.97	1.466	265
Di-isopropylether	0.28	3.88	0.37	1.368	220
Toluene	0.29	2.38	0.59	1.496	290
n-Propyl chloride	0.30	7.7	0.35	1.389	225
Benzene	0.32	2.28	0.65	1.501	290
Ethyl bromide	0.37	9.34	0.39	1.421	230
Diethyl ether	0.38	4.33	0.23	1.353	220
Chloroform	0.40	4.8	0.57	1.443	250
Methylene chloride	0.42	8.93	0.44	1.424	250
Tetrahydrofuran	0.45	7.58	0.46	1.407	220
Dichloroethane	0.49	10.7	0.79	1.445	230
Methylethyl ketone	0.51	18.5	0.4	1.379	330
Acetone	0.56	21.4	0.32	1.359	330
Dioxane	0.56	2.21	1.54	1.422	220
Ethyl acetate	0.58	6.11	0.45	1.370	260
Methyl acetate	0.60	6.68	0.37	1.362	260
Nitromethane	0.64	35.9	0.65	1.382	380
Acetonitrile	0.65	37.5	0.37	1.344	210
Pyridine	0.71	12.4	0.94	1.510	310
n-Propanol	0.82	21.8	2.3	1.38	200
Ethanol	0.88	25.8	1.2	1.361	200
Methanol	0.95	33.6	0.6	1.329	200
Ethylene glycol	1.11	37.7	19.9	1.427	200
Water	very large	80.4	1.00	1.333	180
Formamide	very large	110	3.76	1.448	
Acetic acid	very large	6.1	1.26	1.372	

[a] The values given for the solvent strengths are for alumina

The samples must be soluble in the eluent. This plays a less important role for analytical applications, where very small sample quantities suffice, than for the preparative purposes where complications may ensue.

The components may only be soluble in eluents from which they are weakly retained on a given adsorbent (low solubility and high elution strength). In such cases the system, and hence the stationary phasc, must be changed. Either a reversed-phase system may be utilized or the separation may be attempted via partition chromatography.

Especially for preparative work, the solvent should be easily and quantitatively removable.

In critical cases the eluent viscosity can be another criterion for eluent selection, because the smaller it is the lower is the required pressure drop to achieve a given flow velocity. If two eluents or eluent mixtures of the same "polarity" are available, preference should be given to the one with the lower viscosity.

Of course, the eluent should not interact irreversible with either the sample or adsorbent. Thus, in using acetone or other ketones it must be borne in mind that these compounds may undergo condensation reactions on active adsorbents such as alumina, which change their elution behavior.

1. Eluotropic Series

The empirical arrangement of solvents in order of increasing elution strength is called an *eluotropic series*. Such series are established by determining the retention times with various solvents for a particular substance on a given adsorbent. The shorter the sample retention time, the higher is the "polarity" of the solvent. With the exception of slight differences, the same order has been found for all oxide adsorbents (e.g., Al_2O_3, silica, etc.). Table VI.2 presents an arrangement according to Snyder [1], which differs only insignificantly from the first eluotropic series by Trappe [23]. The sequence is always arranged in order of decreasing sample adsorption (i.e., longest retention times and highest k' values first). The lower the position of the eluent in the series, the shorter are the retention times until, finally, the elution strength ("polarity") becomes so high that the sample is no longer retained. It can be generally stated that the solvent adsorbed more strongly on a given stationary phase will elute more strongly from it. Since the eluent is always present in great excess compared to the sample and competes for the active surface sites, even relatively nonpolar solvents can be used to elute more polar substances.

Table VI.2 summarizes the most important organic solvents. The order is in accordance with increasing ε^0 values ("solvent strength" parameter). According to Snyder [1], these ε^0 values are determined by measuring the retention volumes of standard samples with different eluents but are always related to n-pentane ($\varepsilon^0 = 0.00$) as reference. The logarithms of the quotient of the k' values of the same sample in different eluents is proportional to the difference in ε^0 of the two eluents, provided the adsorbent properties of the column packing are kept constant.

The numerical values of ε^0 (determined on alumina) should only serve to clarify the differences. The ε^0 values for silica fall into the same sequence but are somewhat smaller (i.e., 0.32 for dichloromethane, 0.55 for acetonitrile, 0.75 for methanol). This sequence is similar to that of increasing dielectric constant (DC), which are also included in this table. The differences that appear in some cases may be attributed to selective or specific interactions between the individual eluents and the adsorbents. In addition, traces of polar impurities may completely change the position of the solvent in the eluotropic series. This is certainly the reason for the differences in the order of some important eluents (e.g., methylene chloride, chloroform, diethyl ether) in the various eluotropic series.

The number of eluents suitable for HPLC with a UV detector is considerably restricted. For work below 254 nm the restrictions become much more severe. Generally, in addition to the aliphatic hydrocarbons, only haloalkanes (e.g., chloropropane, dichloromethane, chloroform), ethers (diethyl ether, dioxane and tetrahydrofuran), alcohols, acetonitrile, water and their mixtures can be used. The lowest applicable wavelength for each eluent is also shown in Table VI.2. For some eluents the wavelength given can only be utilized if spectroquality solvents are employed - not a very inexpensive way of carrying out LC.

Perfluorinated aliphatic hydrocarbons (i.e., Fluorinert, 3 M) are much weaker eluents than n-pentane (negative ε^0 values). With such eluents benzene exhibits a k' value > 5. Hence they are useful for separating aliphatic hydrocarbons, olefins, etc. However, their limited solubility for the higher members of these homologous series and their high cost diminish their value as eluents. On the other hand, the separations that can be carried out with this eluent can also easily be performed by gas chromatography.

2. Choice of Eluent

Even today a trial and error approach is necessary for finding a suitable eluent for a completely unknown mixture. Results from thin-layer chromatography (TLC) can be transferred to column chromatography with certain restrictions. If some information about the sample composition is at hand, some general rules (see Chapter XI) may help in the proper selection of the eluent. For completely unknown mixtures it is best to start with a moderately polar eluent such as methylene chloride. If the k' values are too small or even zero, the polarity of the eluent is too high and a less polar one, i.e., one higher in the eluotropic series, must be selected. Another possibility is to reduce the water concentration in the eluent, thereby raising the activity of the adsorbent and increasing the k' values. This alternative should be employed with care because it may be accompanied by a more pronounced isotherm nonlinearity that gives rise to tailing.

If the samples are excessively retained, the elution strength (or polarity) of the eluent must be increased. This can be accomplished by using a solvent that is lower in the eluotropic series than the one tried. Excessive retention times can also be shortened by raising the water content of the eluent (lowering the adsorbent activity). Adjusting the eluent in this fashion is very tedious because of the long time required to re-equilibrate a separation system after a solvent change.

In order to achieve rapid analyses, the elution strength of a solvent should be adjusted so that the k' values are less than 5. Only in the rarest cases can this optimum be attained with a pure eluent, and therefore mixtures of several polar solvents must generally be used because the polarity differences of pure eluents applicable in HPLC are too large.

3. Solvent Mixtures

Only by using solvent mixtures as eluents is it possible to fully exploit the potential of adsorption chromatography. It should be noted, however, that many physical properties, such as viscosity, solubility, and elution strength, do not exhibit a linear relationship with the eluent composition. Snyder [1] has determined experimentally the variation of the elution strength as a function of the composition of several mixtures (Fig.VI.7). The curves have a characteristic shape. Even

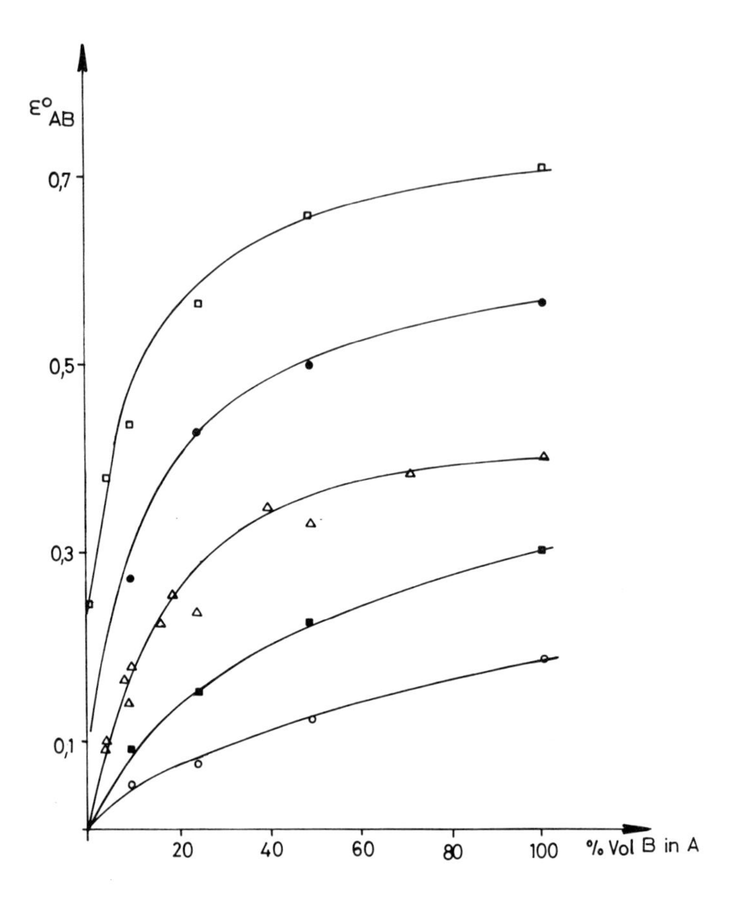

Fig.VI.7. Solvent strength of eluent mixtures (according to Snyder [1]) on alumina. o Pentane-carbon tetrachloride ■ pentane-n-propyl chloride Δ pentane-methylene chloride ● pentane-acetone □ pentane-pyridine

small amounts of the polar component are sufficient to raise the elution capacity considerably. This is particularly striking for mixtures composed of solvents having very different elution powers (e.g., pentane-pyridine or pentane-acetone). The addition of even a few per cent of the polar component causes the elution power to rise rapidly, whereas the subsequent increase is relatively slow. For small differences in the elution strength a nearly linear variation in effectiveness is observed as a function of the composition (pentane/carbon tetrachloride and pentane/n-propyl chloride).

According to Fig.VI.7 the same elution strength ε^0 can be generated by the addition of either a small quantity of very polar sec-

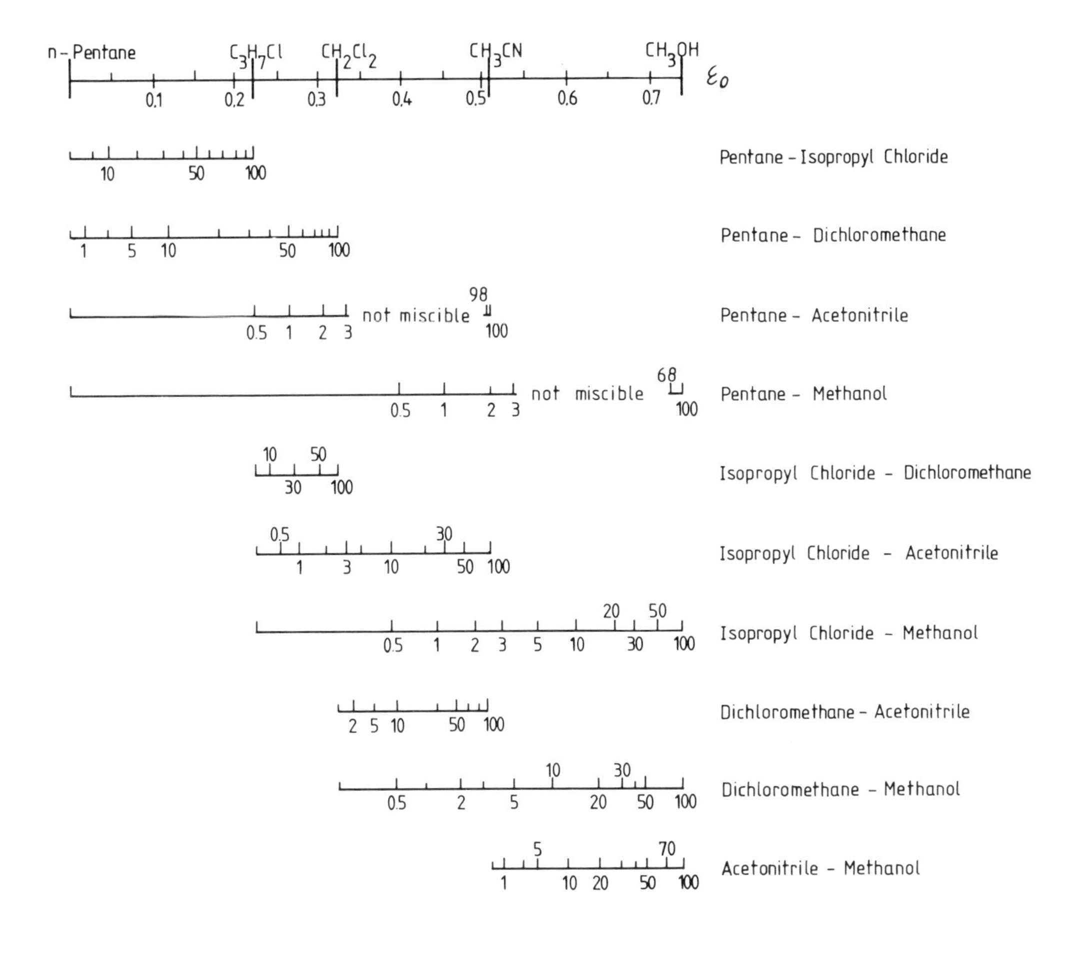

Fig.VI.8. Solvent strength of eluent mixtures for HPLC (similar to Saunders [25])

ond component (e.g., pyridine) to n-pentane or by adding a larger quantity (> 20 v/v%) of less polar solvent (e.g., dichloromethane). The first case (concentrations below 2% v/v) involves all the problems common to moderated eluents, as discussed previously.

The eluent compositions of equal eluotropic strength of the five eluents most frequently used in HPLC with polar stationary phases are depicted in Fig.VI.8 in a manner similar to that of Neher [24] and Saunders [25]. Pentane may be substituted for hexane or heptane without changing the eluotropic strength of these mixtures at all. Each line corresponds to a range (0 - 100%) of binary mixtures. Only aliphatic hydrocarbons are not miscible with methanol and acetonitrile over the entire range. If an eluotropic strength between pentane and propyl chloride is required, it can be produced by mixing propyl chloride and pentane in five or ten percent steps. The same eluotropic strength can be obtained by adding much smaller amounts of dichloromethane to pentane, e.g., the eluotropic strength of a 1:1 mixture of pentane-propyl chloride can be achieved by adding about 10% dichloromethane to pentane. If acetonitrile or methanol is added to pentane to adjust the same solvent strength, amounts far less than 0.5% of these eluents are needed. Similar mixtures can be prepared for propyl chloride with dichloromethane, or acetonitrile or methanol, etc. Other eluents such as ethyl acetate or diethyl ether [25] can also be included in these mixtures.

One should, however, not be surprised if the absolute and relative retentions for a given sample mixture change distinctly when different eluent mixtures having same eluotropic strength are used. These phenomena can be attributed to "secondary solvent effects" [1] which are sometimes difficult to explain. Most of these secondary solvent effects can be ascribed to specific solute-solvent-adsorbent interactions. Some examples are cited and discussed by Snyder [1,26].

One striking example is the change in elution order of acetonaphthalene and dinitronaphthalene on alumina columns [26]. If a benzene-pentane (1:1) mixture is used as eluent, acetonaphthalene is retarded more strongly than dinitronaphthalene. The relative retention in this case is 2.0. If a pentane-dichloromethane mixture (23% v/v CH_2Cl_2) is used as eluent the k' value of acetonaphthalene is almost the same (5.5) but now dinitronaphthalene is eluted as the second peak, the relative retention now being 1.05. If 0.05 vol% dimethylsulfoxide in pentane is used as eluent, acetonaphthalene is only weakly retarded whereas dinitronaphthalene is eluted later. The selectivity of this system, however, is large (α = 3.5).

One of the reasons for these "secondary effects" may be the demixing of the eluent on the adsorbent, the polar component being preferentially adsorbed. In TLC this leads to the appearance of several fronts [27] having different eluent compositions. In a column the corresponding phenomenon occurs only on the initial wetting of the adsorbent. However, since new eluent is being constantly delivered, the adsorbent continuously removes the polar component until an equilibrium state is attained. The adsorption system thus changes more or less into a partition system (cf. Chapter VII).

However, such tremendeous changes in selectivity may also occur on mixing two eluents with very similar elution strengths, such as dichloromethane and ethyl acetate.

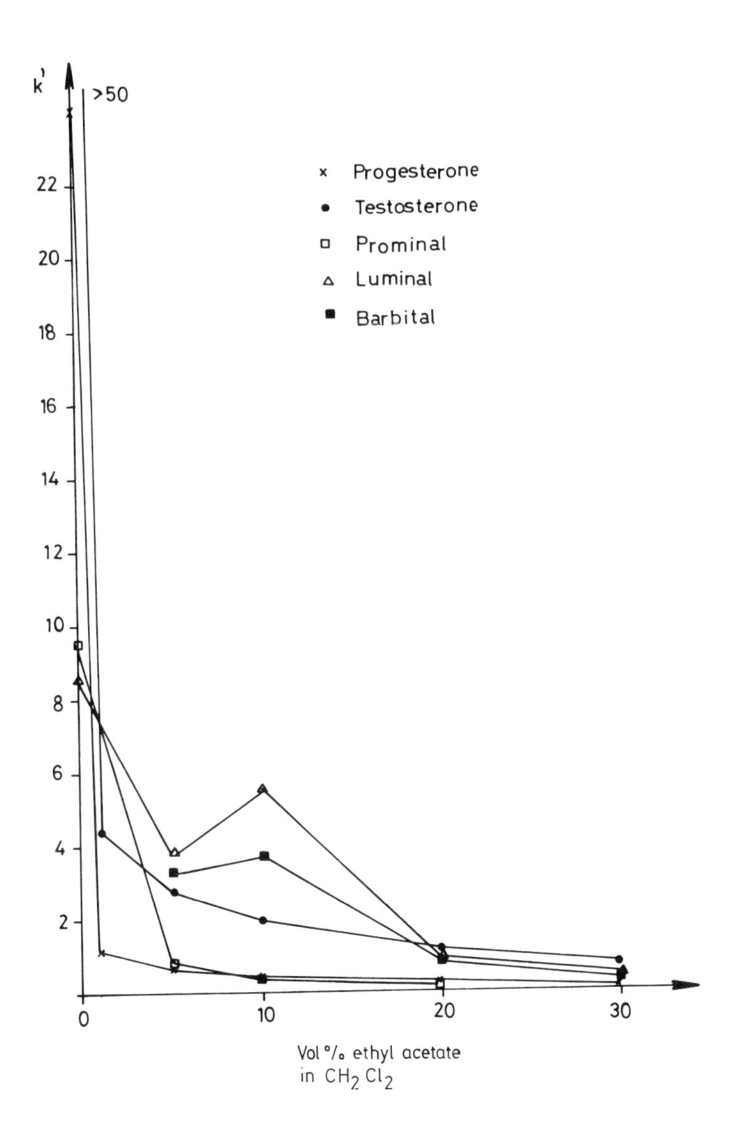

Fig.VI.9. Effect of eluent composition on retention. Stat. phase: "Brush", dinitrophenyl on Merckogel Si 100; eluent: methylene chloride + ethyl acetate

The effect of such eluent composition on the retention behavior of steroids and barbiturates is demonstrated in Fig.VI.9. The k' values of the steroids are reduced drastically by the addition of 1% of ethyl acetate to methylene chloride. Some barbiturates behave similarly, whereas the behavior of others is dominated by other solvent effects (e.g., slight solubility in methylene chloride, good solubility in ethyl acetate, etc.).

For these reasons it is sometimes difficult to reproduce adsorption-chromatographic separations. This is illustrated in Fig.VI.9 using the example of the steroids; In pure methylene chloride the k' value for progesterone is 24, whereas that of testosterone exceeds 50. The addition of 1 vol% dry ethyl acetate to the methylene chloride reduces the k' values to 1.2 and 4.4, respectively. A slight change in the ethyl acetate concentration, for example by evaporation, produces a marked change in the k' values. If, in addition, "secondary solvent effects" also play a role, then both the retention times and the relative retentions of the individual peaks may change (cf. the barbiturates in Fig.VI.9), so that the elution order may even be reversed. Therefore, mixtures should be prepared from solvents having only small differences in elution strengths, for which there is no concern about demixing on the adsorbent and no pronounced dependence of the k' values on minor changes in the eluent composition. On the other hand, the selectivity of a system usually is maximized by utilizing concentrations of less than 3 vol%. The price is reduced reproducibility mainly due to all the problems involved in so-called "moderated" systems [22].

E. Effect of Sample Structure

The molecular structure of the samples determines the elution order to a greater extent than the properties of the solid stationary phase and the eluent. Knowledge of the composition of a sample and the structure of its components simplifies the choice of the system and enables predictions to be made about the elution order.

The strength of retention is almost exclusively a function of the type and number of functional groups. Because the sample concentrations are usually low (< 0.1%), solubility seldom plays a role.

In the simplest case every single atom or group of the sample molecule contributes to the adsorption by specific interactions with the surface. The functional group must be able to interact with the adsorbent surface, i.e., there should be no steric hindrance.

If the functional groups of a given molecule interact with each other (i.e., steric hindrance, resonance, hydrogen bonding, etc.), the strength and type of interaction with the adsorbent surface is altered. The retention behavior of o- and p-nitrophenol is used as the standard example. Because of the intramolecular hydrogen bonding, the ortho isomer has a substantially shorter retention time than the para.

The nature of the functional group determines the elution order. In the following empirical sequence the retention of a compound R-X (where R is the organic moiety and X the functional group) increases in the order:

Alkyls < halogens (F < Cl < Br < I) < ethers < nitro compounds < nitriles < tert. amines < esters < ketones < aldehydes < alcohols < phenols < prim. amines < amides < carboxylic acids < sulfonic acids

Within this series transpositions may occur, depending on whether the functional group is bonded to an aliphatic or aromatic residue. For example, if resonance with the benzene ring increases the charge density on the functional group, the interaction between the stronger "basic" group with the "acid" adsorption surface is enhanced considerably. The hyperconjugation of the alkyl side chain in toluene or ethylbenzene with the benzene ring is responsible for the stronger retention of these compounds compared to benzene, for the alkyl groups themselves do not contribute appreciably to the retention.

To a first approximation, samples containing one functional group, such as a nitro group or one listed before it, can be eluted with hydrocarbons. Aldehydes, ketones and esters can be eluted with dichloromethane. To elute the other compounds methanol or acetonitrile must be added. If the functional group increases the water solubility of the sample, the use of reversed phase systems is recommended.

The strength of retention is increased proportionately by the introduction of a second functional group, provided it does not interact with the first (e.g., the nitrophenols above). In aromatic compounds these groups may enhance or diminish their influence due to resonance effects.

Naturally, the retention strength depends also on steric effects. For maximum interaction the adsorbed molecule must orient itself parallel to the adsorption surface. Bulky alkyl groups adjacent to the functional group diminish the retention. Cis compounds are always retained more strongly than trans (classical example - separation of cis- and trans-azobenzene). Functional groups in the equatorial positions of cyclohexane derivatives and steroids give rise to a stronger sorption than the same groups in the axial position.

Furthermore, the strength of adsorption increases with the size of the molecule (molar volume) especially in nonpolar systems. The effect of these dispersion forces decreases with increasing polarity of the eluent.

In adsorption chromatography a group separation based on functional groups is obtained, with the members of a homologous series being eluted very closely together. In the aliphatic series, increasing the chain length scarcely affects retention, especially if the number of methylene groups exceeds 4 - 6. Only the lower homologs can be readily separated by adsorption chromatography, whereas the higher homologs (more than 10) cannot be separated at all. In such a case separation by reversed phase or partition chromatography would be more fruitful.

It is beyond the scope of this book to discuss all of the effects of sample structure on retention. The literature [1] should be consulted for a detailed discussion. In summary, it will only be repeated that: the strength of adsorption depends significantly on the type of functional group and the ability of this group to approach the solid surface.

II. Nonpolar Stationary Phases

A. General

Since the introduction of chemically bonded phases, the preparation of nonpolar stationary phases has become simple and reproducible. They are also called reversed phases (RP) because in contrast to normal chromatography, the stationary phase is nonpolar (hydrophobic) and the strongest sorption (highest retention) occurs from the most polar eluent - water. The retention can be decreased by reducing the eluent polarity, for example, by the addition of methanol. Reversed phase chromatography dates back to the work of Howard and Martin [28], who coated kieselgur with paraffin oil and n-octane, and utilized this partition system to separate fatty acids. The coatability of kieselgur with nonpolar stationary phases was subsequently improved by silanizing with dimethyldichlorosilane. The introduction of the bonded phases has rapidly led to significant developments in separation technology, as evidenced by the fact that in 1974 only about 20%

of the HPLC separations were carried out on RP [29], whereas recently this level has risen to 60 to 80%.

Chromatographic behavior similar to that of an RP is also shown by active charcoals, especially if they have been graphitized by high-temperature treatment. Their mechanical stability is poor [30], but can be improved by thermal treatment [31]. Stable active charcoals can be prepared by thermal decomposition of benzene on silica gel [31]. Reversed phases made from carbon are not yet commercially available, and therefore only the properties of RP prepared by reacting silica gel with alkylsilanes will be described here. Since the properties and advantages of chemically bonded phases have already been discussed (cf. Chapter V,B) and reviewed [33], only the very specific properties of RP will be considered here.

B. Reversed Phase Properties

Reversed phases can be prepared fairly simply by reacting silica gel with mono-, di-, or trichloroalkylsilanes (cf. Chapter V,B). Various stationary phases are commercially available, with chain lengths ranging from C_1 to C_{18}. The designation of these phases is generally based on the chain length of the longest alkyl group bonded to the support surface. The one or two methyl groups introduced concurrently (depending on whether di- or monochlorosilanes are used in the preparation) exert little influence on solute retention and are usually neglected, except in the cases of the bonded dimethyldichlorosilane phase which is commercially available as RP 2.

The RP should be tested for the presence of hydroxyl groups. In "true" RP, all surface silanol groups should be reacted or shielded. They should no longer adsorb methyl red (cf. p. 91). A far more sensitive test for unreacted silanol groups [34] is the retention of polar samples from nonpolar eluents, such as n-heptane. The retention on silica gel from nonpolar eluents is known to depend on the interaction of the sample with the silanol group [1]. The fewer accessible silanol groups, i.e., the more that are substituted or shielded by alkyl groups, the lower is the retention of polar compounds from nonpolar eluents. On an RP with the silanol groups completely substituted, all such sample peaks should coincide with the unretained peak. On a "good" RP, for which the methyl red adsorption test is negative, nitrobenzene should exhibit a k' value of less than 0.5 when eluted with n-heptane. (The corresponding value on bare silica gel is greater than 10.)

Another means of characterization is the determination of the quantity of bonded organic stationary phase, such as by C, H analysis. However, this provides no information regarding the accessibility or shielding of the silanol groups. Assuming octadecylsilane is used, between 16 and 22 w/w% carbon (based on C, H analysis) can be chemically bonded on silica gel whose specific surface area is 300 - 400 m^2/g (e.g., Lichrosorb Si 100). With alkylsilanes of a shorter chain length the amount of bonded organic phase is, of course, smaller. In addition, the amount of bonded carbon depends also on the pore structure (the pore size and specific surface area). Thus, on a silica gel with a specific surface area of 50 m^2/g (a pore size of *ca.* 300 Å) only about 4.5 w/w% of carbon can be bonded [35].

In other words, approximately 3 µmol organic moieties can be bonded per m^2 of specific surface area. This value holds for octyl to octadecyl groups on silicas with an average pore diameter > 60 Å. Increased surface coverage is observed with shorter alkyl groups. Because of this, optimum shielding of the silanol groups is attained with propyl or butylsilanes. With longer alkyl groups the extent of reaction with the silica is smaller for steric reasons, but even in this case the silanol groups are extensively shielded.

The amount of carbon bonded to the surface determines the k' values of the samples for a given eluent composition, i.e., the smaller the phase ratio V_m/V_s becomes, the higher are the k' values. For a homologous series of samples, a plot of log k' against the number of carbons in the sample always yields a straight line. The slopes of these lines increase with increasing chain length of the alkyl groups bonded to the surface.

This is illustrated for RP with different alkyl groups (C_4, C_{10}, and C_{18}), using n-alcohols as samples and pure methanol as eluent (Fig.VI.10). Obviously, the greater the slope, the larger is the relative retention of two adjacent members of a homologous series; it is therefore larger on C_{18} RP than on those with shorter alkyl groups [34,36]. For the same bonded alkyl group concentration, the k' values vary directly with the specific surface area of the silica, but the relative retentions are independent of it.

The sparing water-solubility of some samples necessitates a higher proportion of the organic component in the eluent, which, however, reduces the k' values. In order to attain k' values in the de-

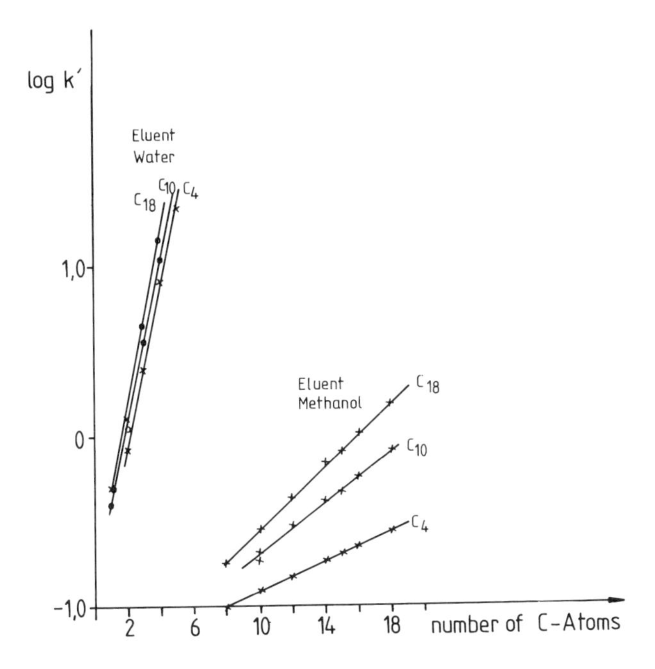

Fig.VI.10. Influence of RP chain length on sample retention with methanol and water eluents. Stationary phases: silica gel Si 100 reacted with butyl-, decyl-, and octadecylsilanes; samples: 1° alcohols

sired range, RP with a high carbon content must be employed. On the other hand, the speed of analysis is always greater on RP with a lower carbon content at constant relative retentions.

These statements appear to be valid, however, only if the water concentration does not exceed 50 - 60% v/v [37,38]. Thus, at higher water contents or in pure water there is virtually no difference in the relative retentions on RP-4 and RP-18 [34] despite the disparity in the carbon contents (cf. Fig.VI.10). Although the absolute retention times do vary, the differences are smaller than expected on the basis of the carbon content and relative to the retention times observed in methanol. Thus, they behave like phases containing less bonded carbon.

The load capacity of the RP C_{18} is an order of magnitude greater than that of bare silica (approximately $2 \cdot 10^{-3}$ g sample/g of stationary phase). This load capacity is a function of the amount of carbon bonded. For a RP C_{18} with ca. 20% w/w bonded carbon it is about double that of a RP C_4 phase (ca. 7% w/w bonded C).

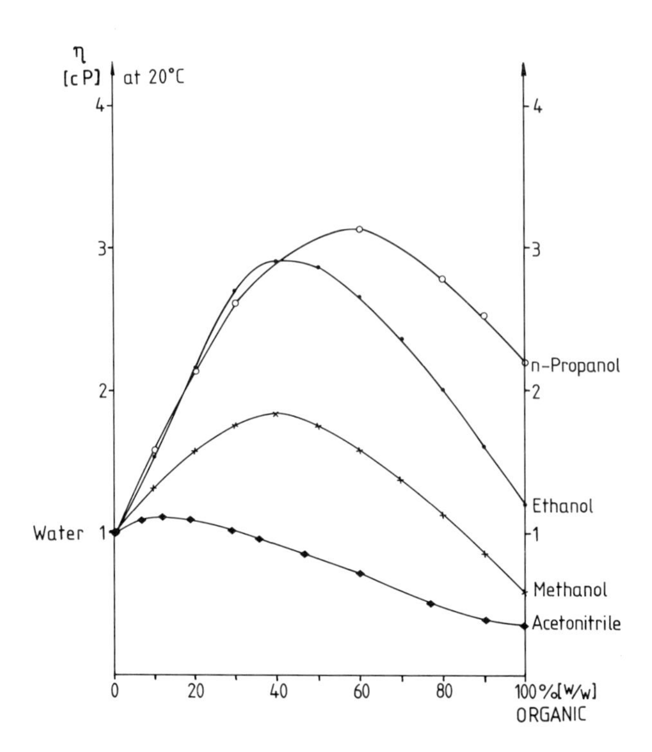

Fig.VI.11. Viscosity of aqueous mixtures of various eluents

The efficiency of RP columns is the same as that of columns packed with silica of the same particle size, under otherwise identical conditions. However, the H values of RP columns are 2 to 3 times larger due to the differences in eluent viscosity and diffusion coefficients. The viscosity of mixtures of water with alcohols and acetonitrile exhibits typical nonideal behavior. The dependence of the viscosity on eluent composition is depicted in Fig.VI.11 [37]. The viscosity for aqueous methanol passes through a maximum of 1.84 cP (at 20°C) at 40% w/w methanol. Therefore, the H values also show a strong dependence on the eluent composition. With water-acetonitrile mixtures the 10% increase in viscosity at about 12% acetonitrile (w/w) is scarcely noticed in routine work. If the phases are not wetted, the H values for retarded compounds are sometimes higher than expected, especially when pure water is used as eluent.

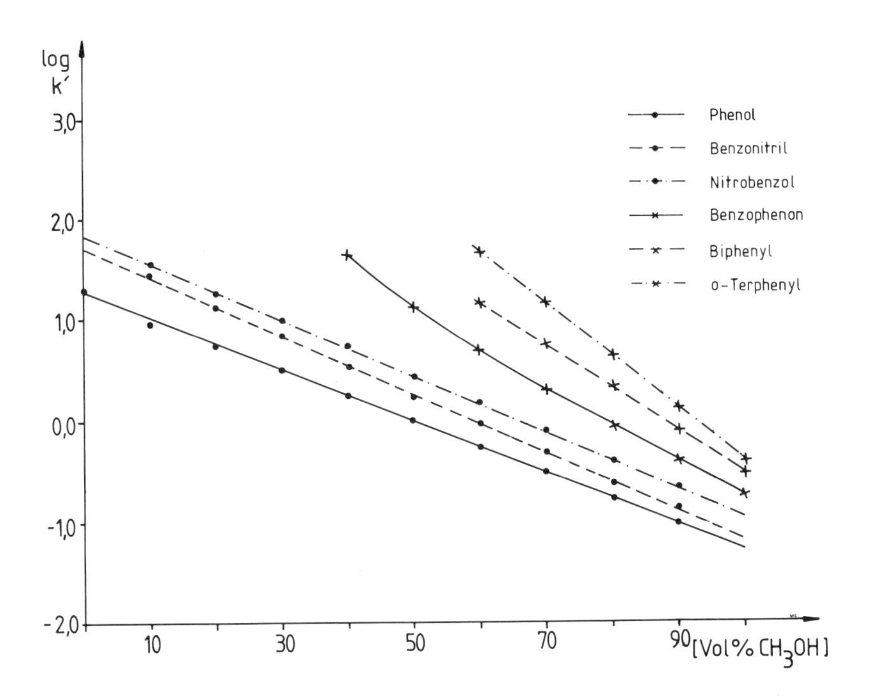

Fig.VI.12. Dependence of k' of simple standards on methanol concentration in water. Stationary phase: silica Si 100 reacted with octadecylsilane

C. Effect of Solvent on Separation

For reversed phases the retentions of organic samples are always largest with water as eluent. Sample elution can be accelerated by increasing the concentration of the organic solvent in water, i.e., by reducing eluent polarity. For water/methanol mixtures a linear dependence of log k' on the alcohol concentration is usually obtained, at least as long as the samples are soluble. This is shown in Fig. VI.12 for some very simple model compounds.

All other organic solvents miscible with water are stronger eluents than methanol. The more strongly a solvent is retarded on an RP from water, the greater is its elution strength alone or in aqueous mixtures. Table VI.3 summarizes the relative retentions, relative to methanol, of the most important UV-transparent, water-miscible solvents, and compares these for C_8- and C_{18}-RP [38]. (The k' value of methanol in water as eluent is around 0.2 on RP C_{18} with D_2O as inert

Table VI.3. "Eluotropic Series" for reversed phases (relative retentions of various eluents relative to methanol in water)

	C_8	C_{18}
Methanol	1.0	1.0
Acetonitrile	3.2	3.1
Ethanol	3.2	3.1
Isopropanol	8.4	8.3
Dimethylformamide	9.4	7.6
n-Propanol	10.8	10.1
Dioxane	12.5	11.7

sample). This order corresponds to a reversal of the eluotropic series for adsorption chromatography (cf. Table VI.2). Consequently, the strongest eluents for RP systems - the aliphatic hydrocarbons - are the weakest eluents on silica gel. For example, if the same concentrations of ethanol or isopropanol in water are used, the samples will be eluted more rapidly with the latter. For relatively nonpolar compounds such as fats (mono, di, triglycerides) nonaqueous eluent mixtures (e.g., acetonitrile - dichloromethane) are required. The same empirical rules as set forth for adsorption chromatography on polar stationary phases are also valid for mixtures of two eluents in this case: It is better to use a higher concentration of a weak eluent in mixtures than to employ a low concentration (a few %) of a strong one. The preparation of the latter mixtures requires greater care and may also lead to the elution of several substances within a single peak due to undesirable displacement effects. The linear relationship between the log k' and the concentration of the organic component in water does not hold even for the acetonitrile-water [39], or the isopropanol-water system. The addition of acetic acid to the eluent also reduces the retention time, similar to that with methanol. In this way the separation of acids may be modified by the addition of acetic acid.

The increase in the k' value resulting from the lengthening of the alkyl chain (e.g., from C_4 to C_{18}) may be compensated by increasing the methanol content of the eluent. This may be desirable, for example, to attain identical retention times on C_4- and C_{18}-RP. As a rule of thumb, a 10% increase in the methanol concentration reduces the k' values on a RP C_{18} to the level obtained on an RP C_4 column.

The addition of neutral salts (Na_2SO_4, etc.) to aqueous eluents leads to an increase of the k' values of hydrophobic samples. Similar procedures are known in organic chemistry as salting out effects.

Dissociated samples may be retained or excluded, i.e., eluted in front of a neutral inert sample. The cause of the latter may be attributed to phenomena similar to the Donnan potential in ion exchange chromatography [42,43]. This effect can be suppressed by the addition of neutral salts (0.1 - 0.3 molar) to the eluent, or by decreasing the dissociation of the samples by adding either acid or base (restricted to pH < 9) to the eluent. If appropriate retention cannot be attained in this way, the use of ion pair chromatography is recommended.

Chemically bonded phases have a considerable advantage over bare silica gel because the conditioning of a column following an eluent change is very rapid. In many cases pumping through at least 10 column volumes of the new eluent suffices to reestablish equilibrium.

The accurate determination of the dead time (t_o) is difficult especially in water as eluent because of the lack of UV-active unretained substances. Hence, recourse to a differential refractometer is necessary, with D_2O serving as the unretained sample. For water-methanol eluents a D_2O-CH_3OH mixture of the same composition should be injected as the unretained substance. This obviates the two principal difficulties that arise when either water or methanol is employed as the inert substance: The CH_3OH is slightly retarded in water or water-rich mixtures and pure H_2O or CH_3OH may give rise to multiple peaks due to so-called vacancy effects.

If only a UV detector is available, finding an unretarded, UV-absorbing substance may pose some difficulties because dissociable unretained substances may even be excluded.

Estimation of the dead time (cf. II.B or III.E.5) is accompanied by large errors for RP systems because chemical modification of silica gel changes its pore volume. Knowledge of the pore volume of the RP or the total porosity ε_T of the column is required for an accurate calculation. From experience, ε_T is 0.75 for RP based on silica gel having a pore diameter of 100 Å. In critical cases, the ε_T of RP columns can be determined by using methylene chloride as eluent and benzene as sample. The silica pore structure or the column porosity are altered insignificantly on changing the eluent from methylene chloride to methanol or to methanol-water (1 : 1) mixtures.

Increasing the column temperature exerts the same effect on retention behavior as increasing the organic component in water, i.e., a reduction in the analysis time. To a first approximation, a 30°C temperature rise halves the retention time for the same eluent composition [39]. However, this also reduces the selectivity of the separation system. It was shown that selectivity and analysis time are identical, when, for example, a separation is performed either at 25°C using a 60 : 40 acetonnitrile-water mixture, or at 60°C with a higher water concentration (53 : 47). At the higher temperature, however, column efficiency is enhanced due to increased mass transfer (reduced eluent viscosity).

D. Effect of Sample Structure

The chromatographic behavior of reversed phases is more easily understood than that of silica gel. To a first approximation, sample retention increases with diminishing solubility in water, i.e., with decreasing polarity. Within a homologous series the retention rises with increasing C-number. For n-alcohols, for example, increasing the sample chain length by a methylene group raises the k' value by 4 units when water is the eluent. In water-organic solvent mixtures the corresponding contribution is smaller.

Samples with branched alkyl groups always have shorter retention times than those with straight chains, the most branched ones eluting first. Hence, for the butyl alcohols the elution sequence is tert-butanol, sec-butanol, iso-butanol, n-butanol; the relative retentions from water-methanol (9 : 1) are greater than 1.1 between all samples [38].

The chromatographic behavior on RP is very similar to the elution order obtained for the GC separation of straight- and branched-chain hydrocarbons on graphitized carbon black [44]; the differences may be attributed to the dispersion forces of the samples. In addition, hydrophobic effects [39] or solvophobic interactions [40,41] have been discussed for the retention mechanism of RP systems.

Fig.VI.13 demonstrates the effect of the alkyl substituents on the retention on relatively large molecules such as the sym. triazines, which are used as herbicides [45]. Samples differing only by one methylene group in the side chain are separated from each other with relative retentions $>$ 1.3 from 75 : 25 methanol-water.

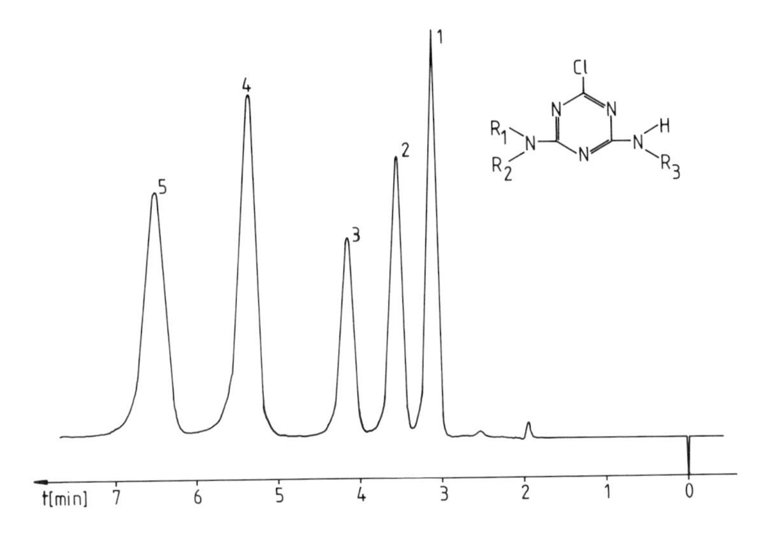

Fig.VI.13. Effect of alkyl substituents on sym. triazines on retention behavior. Stationary phase: Si 100-C_{18}-reversed phase; d_p ~ 10 µm; column 30 cm, 4 mm i.d.; eluent: water-methanol (25:75); F = 1.6 ml/min; u = 2.6 mm/sec; Δp = 130 atm; UV detector: 254 nm. Samples:

1 = Norazin $R_1 = H$; $R_2 = C_3H_7$; $R_3 = CH_3$; k' = 0.60
2 = Atrazin $R_1 = H$; $R_2 = C_3H_7$; $R_3 = C_2H_5$; k' = 0.83
3 = Propazin $R_1 = H$; $R_2 = C_3H_7$; $R_3 = C_3H_7$; k' = 1.14
4 = Trietazin $R_1 = C_2H_5$; $R_2 = C_2H_5$; $R_3 = C_2H_5$; k' = 1.77
5 = Ipazin $R_1 = C_2H_5$; $R_2 = C_2H_5$; $R_3 = C_3H_7$; k' = 2.36

RP systems are primarily applicable to the separation of very polar samples that are only soluble in highly polar eluents such as water, alcohols, etc., from which sorption on silica gel is fairly small. However, it is also entirely feasible to separate aromatic hydrocarbons on RP, using methanol as eluent; it is even possible to separate saturated from unsaturated hydrocarbons. Thus, on a C_{18}-RP using methanol as eluent the relative retentions between adjacent pairs for the series diphenyl, phenyl cyclohexane, and bicyclohexane are greater than 2 [34]. Unsaturated fatty acids are always eluted before the saturated analogs. If silver ions are added to the eluent, the k' values of saturated compounds remain constant, whereas those of unsaturated compounds decrease. The effect increases with increasing degree of unsaturation [46].

The effect of other substituents on the retention behavior is illustrated in Fig.VI.14 for substituted phenols with water as eluent. The introduction of one, two, or three methyl groups into a phenol molecule increases log k' by a fixed amount. The contribution of an

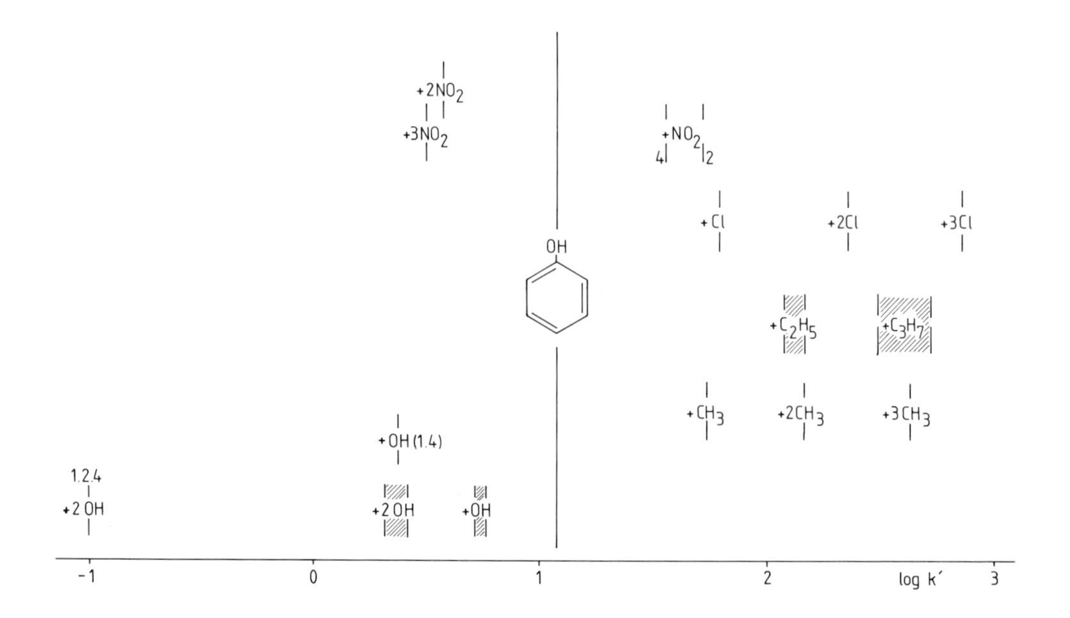

Fig.VI.14. Influence of functional groups on retention behavior. Substituted phenols on RP C_{18} (column: 30 cm, 4 mm i.d.; eluent: water)

ethyl group approximates that of two methyl groups, although the substituent position (ortho- or para to the hydroxyl group) plays a role, of course. The trimethylphenols are eluted together with the propylphenols. The introduction of a chlorine atom into the phenol also raises the log k' values by a constant amount. A similar result is obtained on introducing one nitro group, but a second or third causes the k' values to decline substantially. Picric acid and 2,4-dinitrophenol are readily soluble in water and are eluted before phenol. A second hydroxyl group lowers the k' values, and hydroquinone elutes considerably earlier than pyrocatechol or resorcinol. The trihydroxybenzenes behave similarly.

Similar interpretations can also be made for steroids and other classes of compounds. In these cases too, the elution order can be predicted from the structural formula or the number and nature of substituents. Thus, steroids with a hydroxyl group elute prior to those with a carbonyl group in the same position. Acetyl derivatives are more strongly retarded than the analogous hydroxyl dervatives. The introduction of a double bond also causes the k' values to diminish.

Fig.VI.15 demonstrates the decrease of k' values by introducing a polar functional group onto a benzene molecule. The influence of

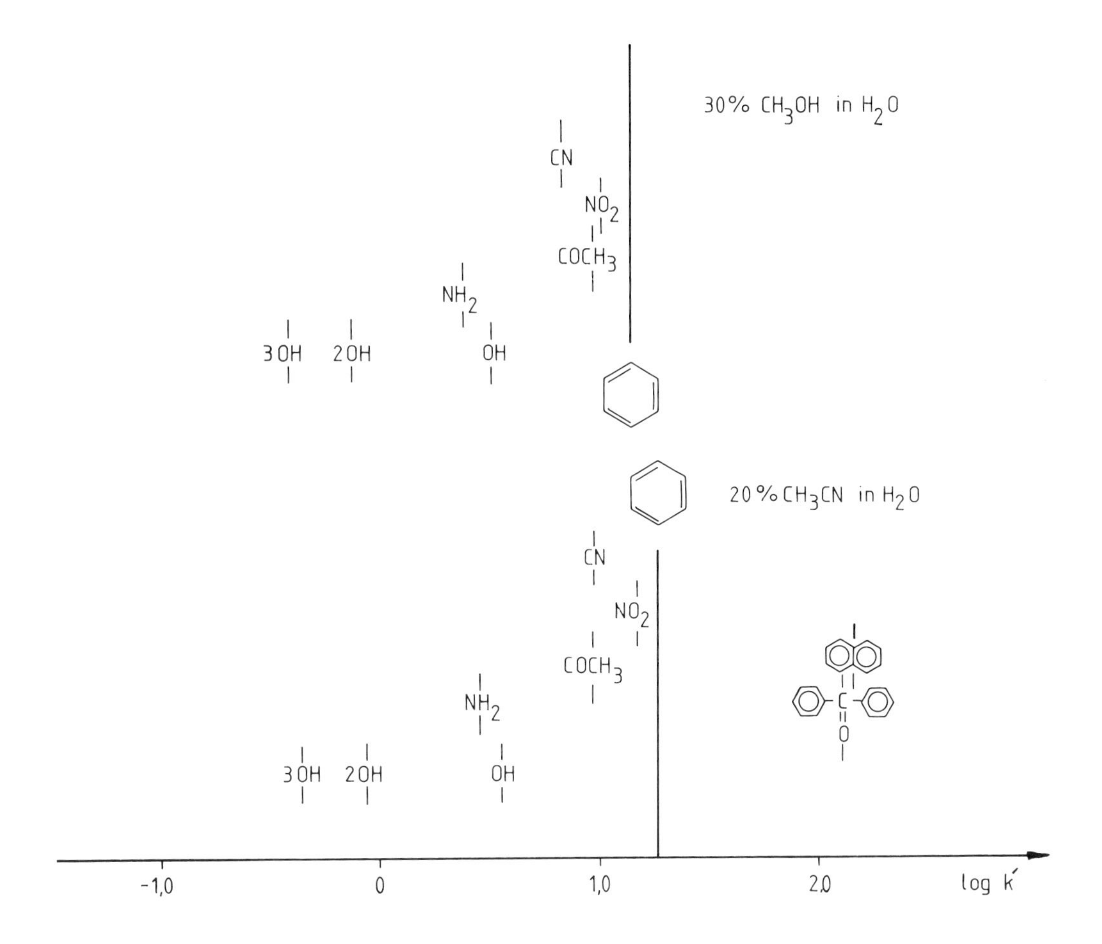

Fig.VI.15. Effect of functional groups on retention behavior. Simple benzene derivates on RP C_{18}. Eluent: 30% aqueous methanol, 20% aqueous acetonitrile

eluent composition is demonstrated also. The eluents are adjusted so that the k' values of benzene are similar in both cases. A 30% methanol concentration in water corresponds to 20% aqueous acetonitrile. The introduction of a polar group into the benzene molecule reduces the k' values. The decrease, however, depends on eluent selectivity. For example, acetophenone cannot be separated from benzonitrile using the acetonitrile - water mixture, but this can easily be achieved with aqueous methanol.

These few examples demonstrate that the retention behavior on RP is clearer and more predictable than on polar phase systems. A linear relationship between log k' and the carbon number is always observed for a homologous series. If some compounds do not fall on this line, they should be classed with another homologous series. An increase in the hydrophobic nature of a sample brought about by a substituent always raises the k' value by an amount that is characteristic of that substituent. Further discussion or theoretical treatment of this hydrophobic or solvophobic contribution is beyond the scope of this introduction.

III. The General Elution Problem

Under *isocratic* conditions, i.e., constant eluent composition and separation conditions (temperature and pressure) a sample mixture may be separated optimally only if the components have k' values less than 10 ($0 < k' < 10$). In some cases, when the more strongly retained components are present at higher concentrations, samples may still be eluted as recognizable peaks even it they have larger k' values. Complex mixtures containing components whose k' values differ considerably cannot be separated and eluted in a reasonable amount of time under isocratic conditions.

To *optimize a separation*, i.e., to resolve both weakly and strongly retained components and to elute these as easily recognizable peaks, use of one of the various programming techniques is necessary. In this way, each sample peak may be eluted under optimal conditions.

The following variables may be programmed in HPLC:

1. Eluent velocity (by programming the inlet pressure)
2. Separation temperature (called temperature programming)
3. Stationary phase
   a) by changing adsorbent activity
   b) by means of coupled columns
4. Eluent composition (called gradient elution)

All of these programming techniques, with the exception of 3 a), can be carried out efficiently with both polar and nonpolar stationary phases.

A theoretical comparison of these methods has been given by Snyder [47]. Only the practical aspects will be discussed here.

The *resolution* of sample zones is always poorer for programmed than for isocratic analyses. The analysis time, however, is shortened and optimized. This follows because the separation of samples with very large k' values always involves a resolution that is greater than necessary. Moreover, the samples are eluted as sharper and, hence, more concentrated zones, which increases the detection sensitivity for the later-eluting peaks. Only the gradients that sharpen elution bands are of practical significance. These include the following:

Increase in the eluent velocity
Increase in the temperature of separation
Decrease in the activity or specific surface area of the sorbent
Increase in the elution strength

Fundamental and technical difficulties increasing in the above order attend the use of these programming techniques in HPLC. However, their separation capability and range of application also increase in the same order.

A. Pressure or Flow Programming [48]

To a good approximation, the retention time is inversely proportional to the pressure drop along a column under otherwise constant conditions. A linear increase in the inlet pressure and, hence, in the flow velocity results in a linear decrease in the retention time. Since the retention times increase exponentially within a homologous series, it is most expedient to employ exponential pressure programming to maintain a constant distance between peaks. It is known from GC that exponential pressure programming corresponds to linear temperature programming [49].

Flow programming may be achieved in the following ways: The output of a pump may be increased continuously by changing the piston stroke. For piston pumps the program shape is affected by the dependence of the output on back pressure. For single-stroke piston pumps the stroke is readily shifted electronically, although the limited output volume poses problems. A high eluent velocity can be achieved very rapidly, but the reservoir volume is depleted in a short time. Pumps with a variable stroke frequency are readily amenable to pressure programming.

No special demands are placed on the stationary phase and the column. The slope (the C term) of the $H = f(u)$ curve should be small

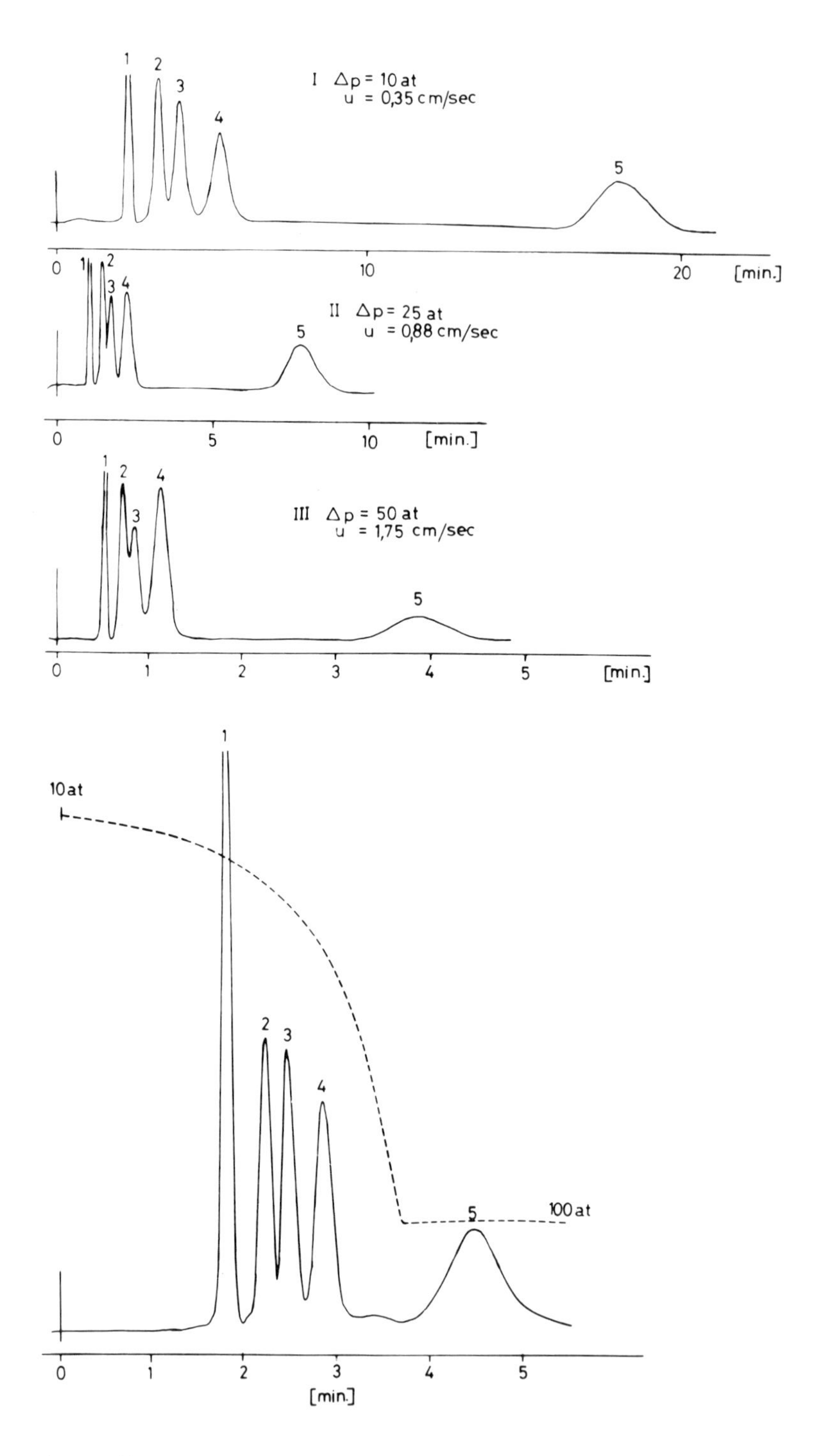

Fig.VI.16. Pressure programming. Separation of insecticides. Stationary phase: silica Si 200; d_p 30 - 40 μm; eluent: n-heptane (~ 30 ppm water); column: 50 cm, 2 mm i.d. I, II, and III isobaric, IV pressure-programmed; samples: 1 = Aldrin; 2 = heptachlor; 3 = DDT; 4 = Lindane

if possible. (Estimation of H values for pressure programmed runs is meaningless because they are defined for a constant linear velocity only). For partition systems, mechanical erosion of the stationary phase is to be expected at high terminal flow velocities (see Chapter VII). Differential refractometers and UV detectors can be readily used for flow-programmed analyses. However, the differential refractometer exhibits baseline drift at programming rates exceeding 6 atm/min due to changes in the flow velocity. The UV detector, as expected, shows no dependence on the flow rate.

The advantages of pressure programming are demonstrated in Fig. VI.16, where isobaric separation at three different inlet pressures is compared to that carried out by pressure-programming. Analysis at the lowest inlet pressure (10 atm) yielded optimal resolution of the first four peaks, but peak 5 required about 20 min to elute. Elevation of the pressure to 25 or 50 atm caused peak 5 to appear in a reasonable length of time, but the resolution of the earlier-eluted peaks deteriorated. An optimum separation with approximately equally good resolution of all peaks was obtained only by pressure-programmed analysis with an exponential pressure rise from 10 to 100 atm.

This example shows that pressure-programmed analysis is most appropriate for chromatograms in which the differences in the sample retention times increase exponentially, i.e., when the resolution of two adjacent peaks is excessive. In such cases it becomes expedient to use an exponential pressure program. If the problem does not involve the separation of the members of a homologous series, any type of program may be employed, and it may be interrupted as often as desired in the course of the run. The primary advantange of pressure programming lies in the ability to return immediately to the initial conditions upon completion of an analysis by depressurizing to the original pressure (in a few seconds).

B. Temperature Programming

The strength of sample retention in adsorption systems depends primarily on the differences in the heat of adsorption between sample and eluent. An elevation of temperature leads to a reduction in the heat of adsorption and, therefore, to a shorter retention time. Hence, temperature programming in LC should give the same results as in GC: The more strongly retained components are eluted more rapidly and as sharper zones at higher temperatures. Unfortunately, the use of temperature programming in LC with polar stationary phases involves fundamental difficulties. An increase in temperature affects the equilibrium distribution of water between the stationary phase and the eluent. As a result, the adsorbed water may be stripped off the column by the eluent. After cooling to the original temperature, the adsorbent will be more active than before, and the sample retention volumes will rise. An increase in temperature should *shorten* the retention times. Depending on the eluent used and the position of water equilibrium, this effect is actually found. For some other eluents an *increase* in the retention times is observed (or they remain unchanged) when the temperature is raised.

These effects were used in temperature programming with the addition of a "moderator" [50,51]. A moderator is a small amount (0.1 - 1%) of a substance (e.g., isopropanol) that is added to the eluent to accelerate sample elution. The moderator is taken up by the eluent from a large pre-column that is also co-programmed. Without a pre-column, "reversed" temperature programs may result from the addition of a moderator.

Therefore, bare silica and alumina are unsuitable as packing materials for temperature programmed analysis. It was shown [52] that the effect of adsorbed water (moderator) is negligible for chemical bonded phases. Hence, temperature programmed analyses can be carried out with these phases without complications. On returning to the initial temperature the column is in equilibrium, and the same retention volumes are observed as before the temperature change. Temperature-programming can be used in RP chromatography to reduce analysis time [53]. However, it is inferior to gradient elution.

Because column packings are poor heat conductors, it is insufficient merely to thermostat and program the column; hence the eluent must be raised to the proper temperature *before* entering the column. For this purpose an approximately 1 m long capillary (i.d. 0.5 - 1 mm) should be installed as heat exchanger and maintained at the same temperature as the column. Although the heating and cooling rates of air thermostats are more favorable than those of liquid thermostats, the former have only a limited applicability for LC and for temperature programming. Liquids possess a considerably higher heat capacity than air, and heat transfer from the air to the eluent is not sufficiently rapid.

Programming with a preheated eluent has the additional advantage of avoiding a radial temperature gradient within the column, which could contribute to band spreading. Since the heat capacities of eluents and column packings are of the same order of magnitude, temperature equilibration between them occurs rapidly.

Only the UV and moving wire detectors are suitable for temperature-programmed analyses.

The effect of temperature on the separation of a mixture of aromatics is illustrated in Fig.VI.17. Isothermal operation at different temperatures is represented in a - c. At room temperature (a) separation of the eight components requires nearly 30 min, whereas at 43°C (b) it is complete in about 12 min, although the first few peaks begin to be somewhat poorly resolved. At 70°C (c) all peaks are eluted after 5 min, but scarcely separated. In contrast, temperature-programmed analysis at 4°C/min (d) yields good resolution of all peaks in about 12 min. A similar decrease in the k' values is usually measured for RP systems. As a rule of thumb an increase of 10% in temperature causes a 10 - 15% reduction in the k' values.

It should be pointed out that the H values decrease with rising temperature. Of course, H values can only be determined under isothermal conditions, never from a temperature-programmed chromatogram. Lower H values are obtained because of the decrease in the eluent viscosity, which is reduced by about 50% for a temperature rise of 50°C. At a constant inlet pressure a decrease in the eluent viscosity during temperature-programming brings about a corresponding increase in the flow rate. This, in effect, is additional flow programming that acts in the same direction: it accelerates the elution of the more strongly retarded components. With constant flow pumps the column inlet pressure drops with increasing temperature but the linear flow rate remains constant.

With respect to the analytical results, there is no difference between flow and temperature programming. Both of these programming modes are appropriate when the resolution of the later-eluting peaks is excessive and wastes analysis time. Which is preferable is difficult to predict. In principle, optimization of the separation should first be attempted by means of flow or pressure programming, as these techniques are simpler and place no special demands on the stationary phase. Such programs do not affect the basis of separation (adsorption, partition, etc.).

If the above approach proves unsuccessful or some peaks elute late with strong tailing, temperature-programming should be tried. This, however, places certain demands on the stability of the stationary phase.

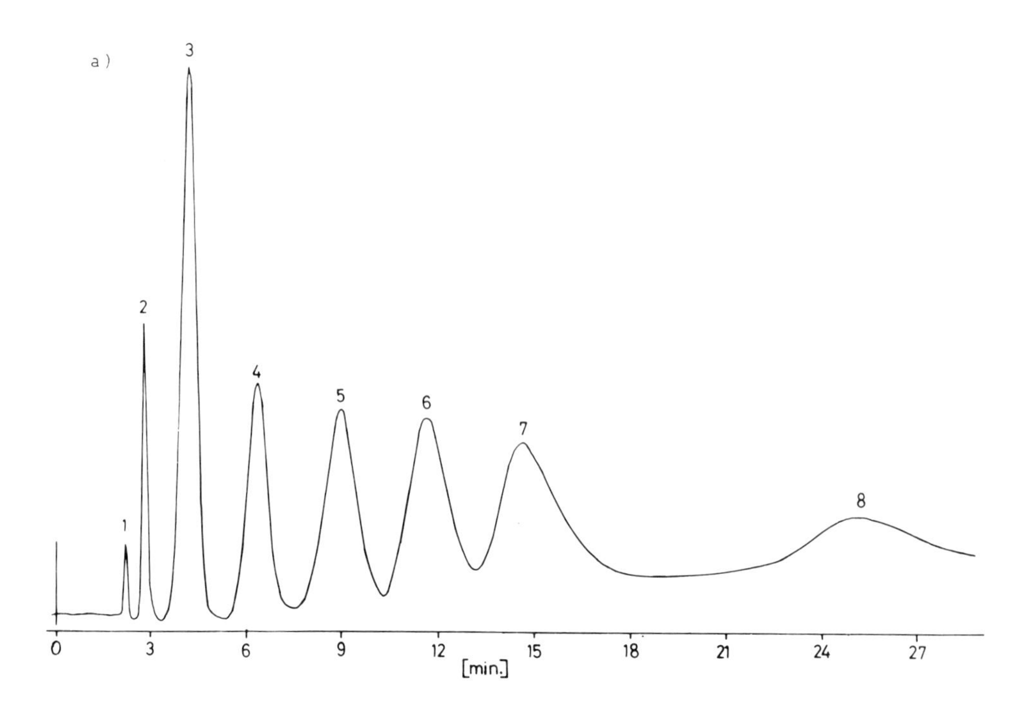
a)
1
2
3
4
5
6
7
8
0
3
6
9
12
15
18
21
24
27
[min.]

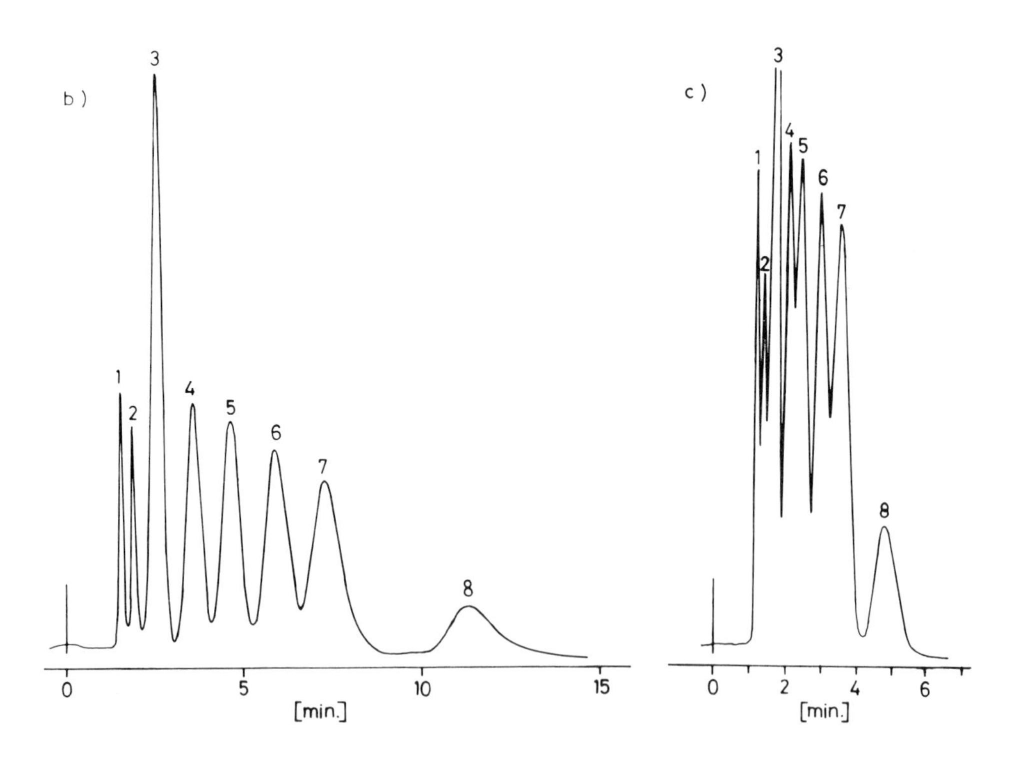
b)
1
2
3
4
5
6
7
8
0
5
10
15
[min.]
c)
1
2
3
4
5
6
7
8
0
2
4
6
[min.]

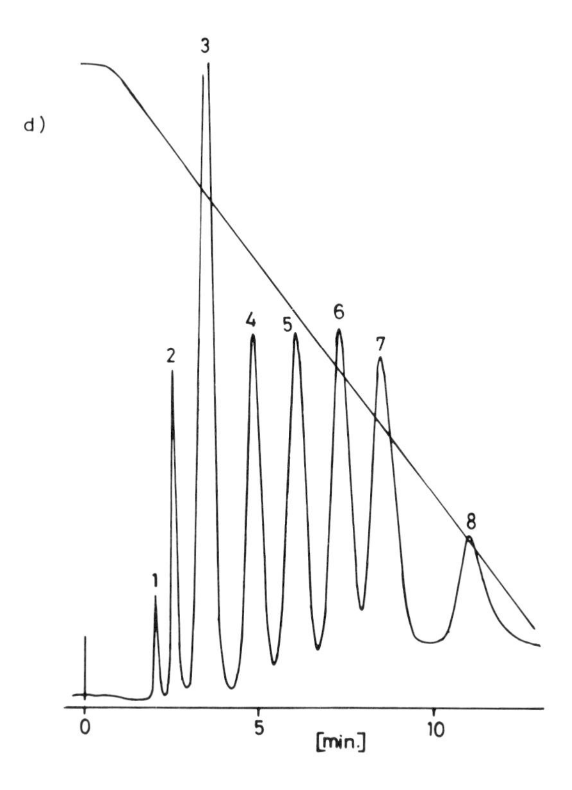

Fig.VI.17. Temperature programming. Stationary phase: "Brush", dinitrophenyl on Si 200; d_p 30 - 40 μm; eluent: n-heptane; column: 50 cm, 2 mm i.d.; Δp = 10 atm. a) Isothermal T = 22°C; b) Isothermal T = 34°C; c) Isothermal T = 70°C; d) temperature-programmed 4°C/min, initial T = 23°C. 1 = unretained; 2 = benzene; 3 = naphthalene; 4 = anthracene; 5 = 2-phenylnaphthalene; 6 = chrysene; 7 = perylene; 8 = picene

Both programming techniques generally lead to a diminution of the k' values by a factor of around 100, i.e., substances that have k' values of 100 to 200 at room temperature elute with k' < 10 when the temperature is raised by 50 to 70°C. These data should be regarded as guideline values because many factors such as eluent velocity, programming rate, column length, etc. play an important secondary role.

C. Programming of the Stationary Phase

1. Variation of Adsorbent Activity

As has already been pointed out, an equilibrium exists between the water dissolved in the eluent and that adsorbed on the active support [10]. As more water is taken up by the adsorbent its activity decreases, and, hence, adsorption of the sample components decreases. A continuous increase in the water content of the eluent decreases the adsorbent activity and accelerates the elution of the polar components. Demixing of the eluent occurs on the adsorbent, i.e., more water is adsorbed at the *front* of the column than at the end. The activity of the stationary phase and therefore the strength of sample adsorption is lower at the beginning than at the end of the column. Such "gradient columns" [54] have already been used in classical column chromatography. The transition region from active to deactivated adsorbent is more or less continuous and moves gradually down the column as more and more water is added. Because of this, the sample components are compressed into sharper zones (a slow migration rate at the top of the column, a faster one at the bottom). In extreme cases the zones are displaced by the water front [55].

The k' values in methylene chloride, for example, are decreased by a factor of $5 \cdot 10^2$ to 10^3 when the water content of the eluent is increased from 60 to 2000 ppm on a silica column. Better resolution is obtained if the water content is increased continuously (as a gradient) rather than discontinuously (stepwise) [53].

The shortening of the elution time is based on programmed deactivation of the stationary phase or on displacement of the components by the very strongly adsorbed water. It must be pointed out that under certain conditions several components may elute as one peak. This phenomenon is observed primarily when a breakthrough of the water zone occurs. In the extreme case the end of a tailing peak can be compressed into a false peak. This problem of *band splitting* has been observed in classical column chromatography [56].

2. Coupled Columns

The coupled column technique is utilized extensively in gas chromatography [57], especially in process GC [58] and has also been used

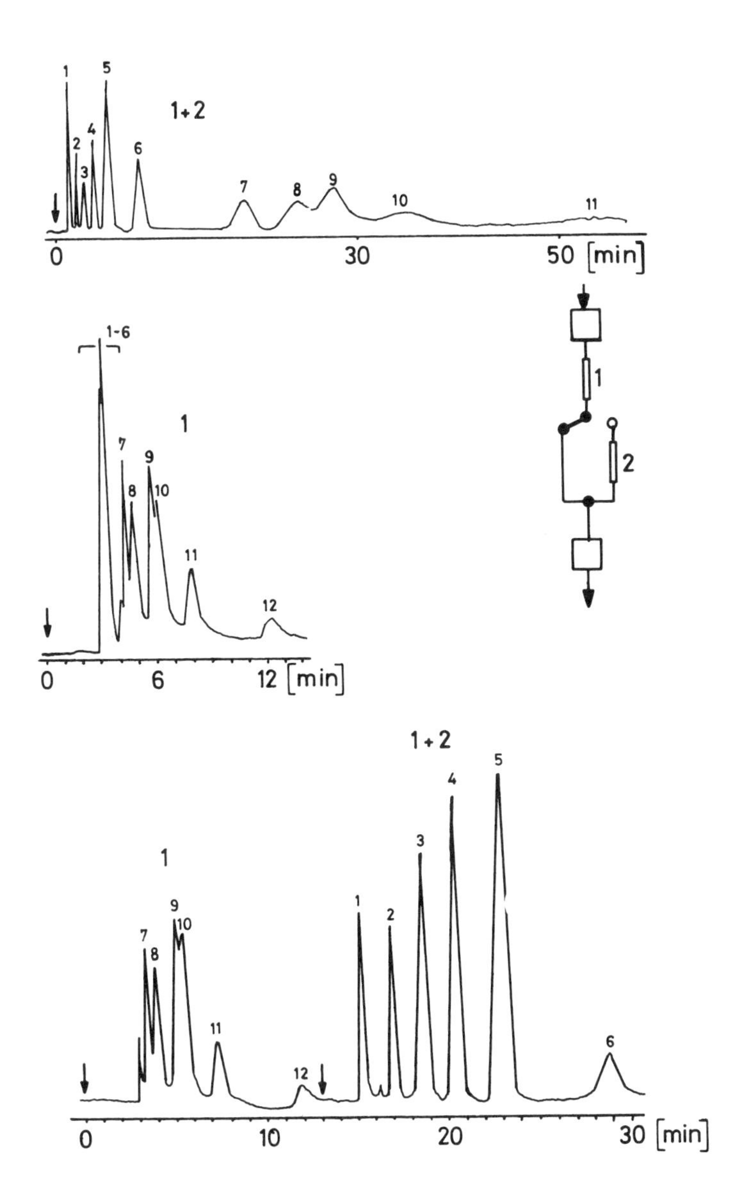

Fig.VI.18. Coupled Columns [61]. Column I: 25 cm, 2.7 mm i.d., kieselgur (2 m^2/g). Column II: 25 cm, 2.7 mm i.d., silica (15 m^2/g). Eluent: water-ethanol-isooctane. Upper chromatogram - columns I and II; middle chromatogram - column I; lower chromatogram - coupled columns, for peaks 7 - 12 only column I, for peaks 1 - 6 column I and II. Samples: 1 = decylbenzene; 2 = progesterone; 3 = androstandione; 4 = methyltestosterone; 5 = testosterone; 6 = androsterone; 7 = 16 α-hydroxy-pregn-4-en-3,20-dione; 8 = 19-hydroxyandrost-4-en-3,17-dione; 9 = corticosterone; 10 = dehydrocorticosterone; 11 = cortisone; 12 = cortisol

in LC [47,59-62]. The mobile phase composition is kept constant, and the sample components are separated on columns containing different stationary phases connected in series or in parallel. Since a constant eluent composition is essential, the separations are based primarily on variation in the column length (number of theoretical plates), the amount of adsorption (the magnitude of the specific surface area), the phase ratio (variable coverage with the same stationary phase), and the selectivity (different chemically bonded stationary phases).

Usually, a preliminary separation is carried out on a relatively short column and then fractions of this eluate are subjected to further separation on other columns containing the same or other stationary phases. Fig.VI.18 illustrates the optimization of the separation of steroids using coupled columns [61]. The steroids that elute first from column I with little resolution are held in column II while the more strongly retained steroids emerge resolved from column I (peaks 7 - 12). After switching the eluent flow to column II, the other steroids (peaks 1 - 6) emerge well separated. In this process the eluent composition remains unchanged. Because of the low diffusion coefficients no appreciable band broadening is observed. Of course, the switching valve and its connections should not contribute additional band spreading.

D. Gradient Elution. Programming of the Eluent Composition

Gradient elution refers to a programmed increase in the elution strength of the mobile phase. It provides the greatest feasibility of all programming techniques for the optimum separation of very complex mixtures. Due to the large variety of possible eluents, mixtures with wide-ranging polarities can be separated on both polar and nonpolar stationary phases. The separation of a mixture containing everything from nonpolar hydrocarbons (e.g., squalane) to water-soluble and polar compounds (e.g., glucose) is demonstrated in Fig.VI.19 by the method of incremental gradient elution in which several (up to 12) gradient steps are used. The eluent combinations were selected by Scott [63-65] on the basis of theoretical considerations so that the resulting gradient increases linearly in elution strength. This eluent series can only be used in conjunction with a wire detector because some of the solvents are not transparent in the UV.

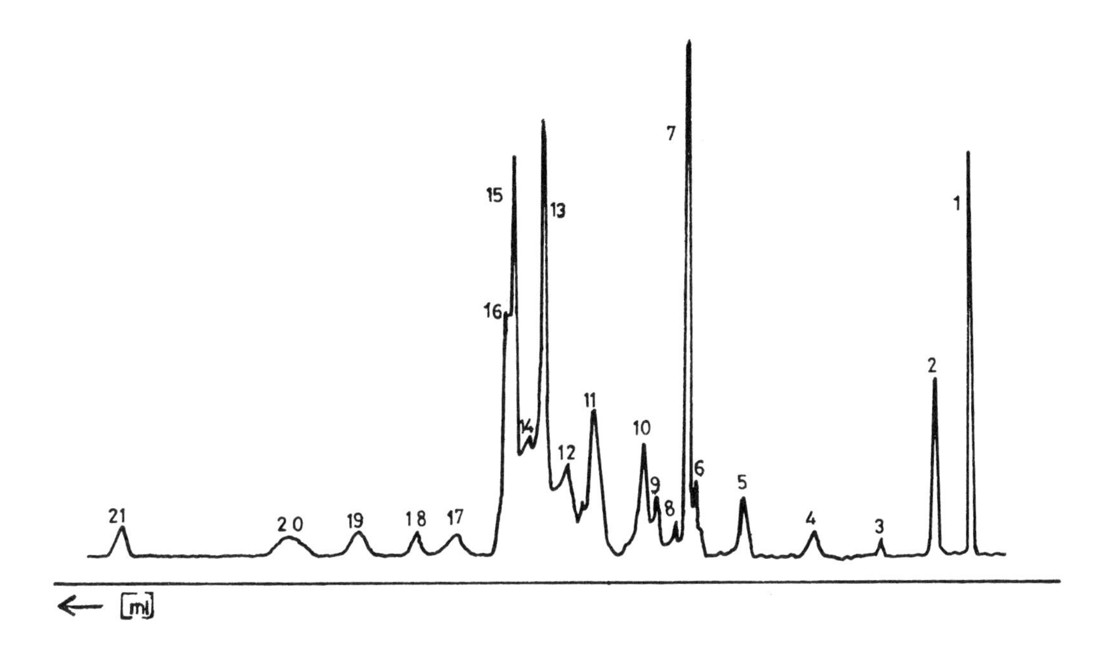

Fig.VI.19. Illustration of gradient elution [63]. Silica, Bio-Sil A; 12 different eluents; column: 50 cm, 2 mm i.d.; F = 0.5 ml/min. Probable peak sequence: 1 = squalane; 2 = anthracene; 3 = methyl stearate; 4 = benzophenone; 5 = chloroaniline; 6 = nitroaniline; 7 = p-dinitrobenzene; 8 = nitrophenol; 9 = dihydrocholesterol; 10 = catechol; 11 = phenacetin; 12 = adenine; 13 = phenolphthalein; 14 = EEDQ; 15 = quinine; 16 = acetylsylicylic acid; 17 = benzoic acid; 18 = BOC-leucine; 19 = BOC-glycine; 20 = alanine; 21 = glucose

Snyder [66] described an eluent series (Table VI.4) that covers a similar range, but consists only of solvents that are completely UV-transparent. It should be noted, however, that sensitive UV detectors also respond to changes in the refractive index, so that at least baseline shifts must be contended with in using this series. Moreover, eluent demixing may occur with this series, resulting in selectivity changes.

The members of the eluent series in Table VI.4 were selected so that the difference in elution strength ε^0 (cf. VI.D.1) between each mixture is 0.05 units. According to Snyder [66], this corresponds to a 2 - 4 fold decrease in the sample k' values.

Excessively large polarity differences are indicated by very large peaks on a chromatogram when the second component breaks through. These peaks result from the simultaneous elution of several unresolved compounds. Fig.VI.20 shows a multi-step gradient elution from heptane to dichloromethane and then further to dichloromethane-isopropanol mixtures. Shortly after changing the elution mixture (which delays the

Table VI.4. Eluent series for gradient elution according to Snyder [66] for silica gel

				Elution strength ε^o
1	Pentane			0.00
2	Pentane/2-chlorpropane	95.8/4.2	(v/v)	0.05
3	Pentane/2-chlorpropane	90/10	(v/v)	0.10
4	Pentane/2-chlorpropane	79/21	(v/v)	0.15
5	Pentane/diethyl ether	96/4	(v/v)	0.20
6	Pentane/diethyl ether	89/11	(v/v)	0.25
7	Pentane/diethyl ether	77/23	(v/v)	0.30
8	Pentane/diethyl ether	44/56	(v/v)	0.35
9	Diethyl ether/methanol	98/2	(v/v)	0.40
10	Diethyl ether/methanol	96/4	(v/v)	0.45
11	Diethyl ether/methanol	92/8	(v/v)	0.50
12	Diethyl ether/methanol	80/20	(v/v)	0.55
13	Diethyl ether/methanol	50/50	(v/v)	0.60

gradient briefly) large peaks emerge, especially when the mixture contains isopropanol. These zones contain several substances and may be further separated via suitable systems.

Practically, incremental gradient elution can only be performed with instruments that permit gradient preparation on the low-pressure side. This requires continuously operating pumps and a small volume between the gradient mixing and column inlet.

Most instruments for HPLC gradient elution permit gradients to be prepared from only two solvents. The various possibilities involved have been discussed in Chapter III.K. The technical problems include viscosity differences of the eluents, variations in compressibility, and nonideal solution behavior. Before a gradient elution is carried out, the reproducibility of the gradient preparation and the constancy of solvent delivery should be checked for the eluent combination chosen.

Two components can be mixed according to various programs. The increase in concentration of the second component as a function of time can be described by concave, linear, or convex curves. A linear gradient curve should be chosen first. The actual, i.e., the effective gradient is only partially dependent on the type of programming.

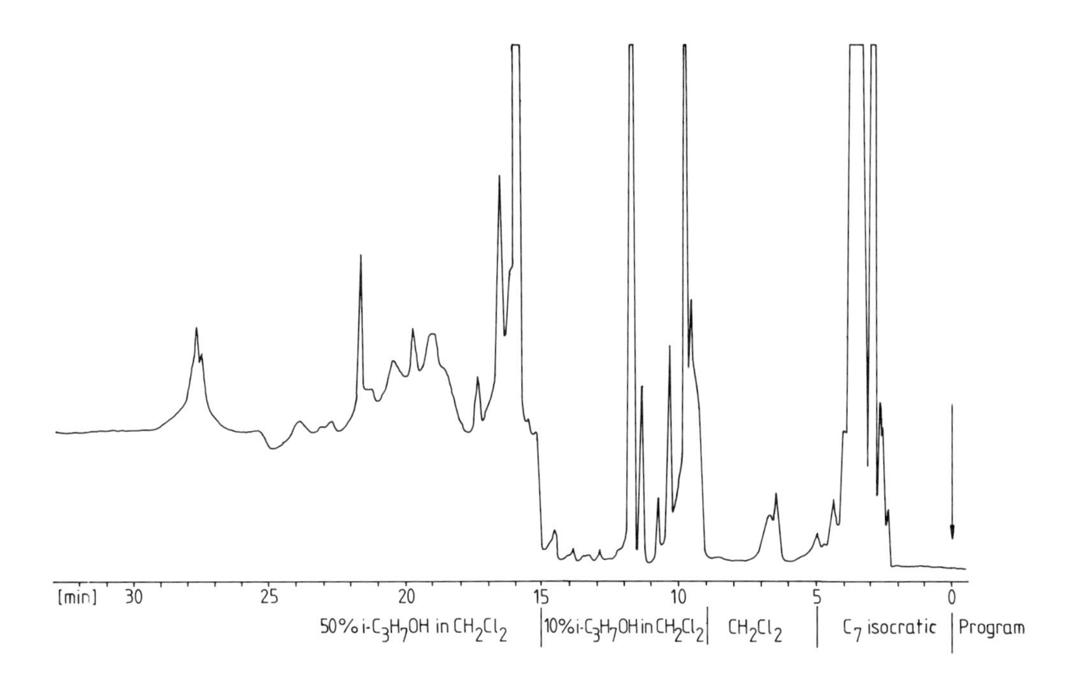

Fig.VI.20. Multi-step gradient elution of oil of peppermint. Low-pressure gradient: column 30 cm, 4.2 mm i.d., packed with silica gel Si 100, d_p ~ 10 µm, F = 2 cm^3/ml. Gradient: mixing chamber volume 10 cm^3 [69]. Eluent program: 1) 5 min = n-heptane isocratically, 2) gradient to methylene chloride, 3) gradient to methylene chloride - 10% isopropanol, 4) gradient to methylene chloride-isopropanol (1 : 1). Column regeneration by flushing prior to sample introduction, a) 15 min with methylene chloride, b) 15 min n-heptane (Uvasol, Merck). UV detector at 254 nm; 0.5 µl sample

In addition, it is always a function of the polarity or elution strength difference of the two components. When two solvents with large differences in their elution strengths are used, even small amounts of the polar component produce a sharp rise in the elution strength, i.e., an effective convex gradient. Such gradient shapes are rarely desirable because the components are almost always eluted close together at the beginning of a separation. Furthermore, displacement effects, similar to those discussed in Section 3a, may appear as a result of demixing of eluent mixtures that consist, for instance, of pure n-hexane and n-hexane containing a few percent alcohol (such as isopropanol).

It has been demonstrated [47,67], that a linear increase in the solvent strength is always optimum for gradient elution, because only then do all peaks show identical peak widths, and the resolution is optimum. Only if the gradient components have relatively small differences in polarity does one obtain an approximately linear increase in elution strength with linear programming. On polar stationary phases this would apply to gradients ranging from n-hexane to the chloroalkanes such as propyl chloride. With a gradient from n-hexane to dichloromethane linear increase in solvent strength can be simulated by using a slightly concave gradient. On nonpolar stationary phases a linear increase in solvent strength is certainly obtained for linearly programmed water-methanol mixtures, and in part also for acetonitrile as the second component.

In gradient elution chromatography the elution volume of a given sample depends primarily on only two variables, if all other conditions, i.e., column, temperature, initial and final eluent composition are kept constant. These are the volume flow rate F (ml/min) and the gradient time t_g. The product of both can be defined as the gradient volume V_g, i.e., the volume of eluent pumped through the system during the programmed run. If the gradient volume is kept constant, flow rate or gradient time can be varied, but the sample components will always emerge at the same eluent composition X_B (%). This was verified experimentally for systems with polar and nonpolar stationary phases, as well as for linear and nonlinear gradient programs [68].

The eluent composition at which the individual sample components elute depends on the gradient volume. In analogy to isocratic chromatography, a capacity ratio for gradient elution k'_g can be defined, (determined from the retention time in the gradient and from the dead time of an inert sample in the initial eluent) which depends on the gradient volume. If the gradient volume is increased the k'_g value increases also, but the samples then emerge at a eluent composition of lower solvent strength, i.e., at a smaller concentration of X_B. At an infinitely large gradient volume the k'_g value of a sample approaches the isocratic k' value in the pure eluent of low eluting strength. If the gradient volume becomes smaller, the k'_g values decrease also and the samples are eluted at a higher concentration X_B. At boundary conditions (very small gradient volumes) the k'_g value approaches the isocratic k' value of the sample in the stronger (second) eluent.

The gradient volume with the usual analytical columns (25 - 30 cm length, 3 - 4 mm i.d.) should not be less than 10 ml, and should be at least 2.5 times the volume of the empty column. With such small gradient volumes the amount of the components that can be separated is, of course, also small. For multicomponent analysis large gradient volumes of about 120 ml should be applied. Gradient volumes of 20 to 40 ml are good values to start optimization of gradient-elution chromatography. The number of peaks that can be separated depends on the flow rate at constant gradient volume. The peak widths (H values are not defined in gradient elution) decrease with decreasing flow rate. The optimum flow rate for 4 mm i.d. columns appears to be between 1 - 2 ml/min (corresponding to linear velocities around 1 - 2 mm/sec). Lower velocities do not reduce the peak volumes appreciably, but prolong the analysis. Moreover, instrumental limitations may cause difficulties in gradient generation.

At constant isocratic efficiency of the column, the peak volume, i.e., the volume of eluent containing the sample on elution, can be changed by varying gradient volume and flow rate. For trace analysis the peak volume can be minimized by reducing the gradient volume and optimizing the flow rate. Very narrow and high peaks are then obtained, but the number of compounds that can be separated, i.e., the peak capacity in gradient elution (quotient of gradient volume and average peak volume) is small. To optimize the peak capacity, the gradient volume should be increased while keeping the flow rate at its optimum value (1 - 2 ml/min). Increasing the flow rate from 1 to 2 ml/min reduces the peak capacity at constant gradient volume by about 10% but cuts analysis time in half.

Gradient elution can also be used for a quick check of the optimum eluent composition for isocratic analysis. To do this, it must be known at which eluent composition during the gradient run the sample emerges from the column. Because of dead volumes in the system, (i.e., in the mixing chamber, connecting tubing, and the volume of the mobile phase in the column), the real eluent composition is always delayed relative to that indicated by the instrument. If the eluent composition is adjusted for this, the isocratic k' values of the samples are around 1 in an eluent composition corresponding to that at which they are eluted in the gradient (this holds for gradient volumes of 20 - 40 ml).

Gradient elution places special demands on solvent purity. Only carefully purified solvents should be used, e.g., it is recommended that they be passed over activated alumina or silica [12]. The column acts as a collector of impurities which may also elute as sharp peaks and be mistaken for sample components. It is therefore advisable to run the gradient alone prior to an analysis to recognize the impurity peaks. Such a blank gradient is shown in Fig.VI.21, and was obtained under the same conditions, but without sample injection, as that in

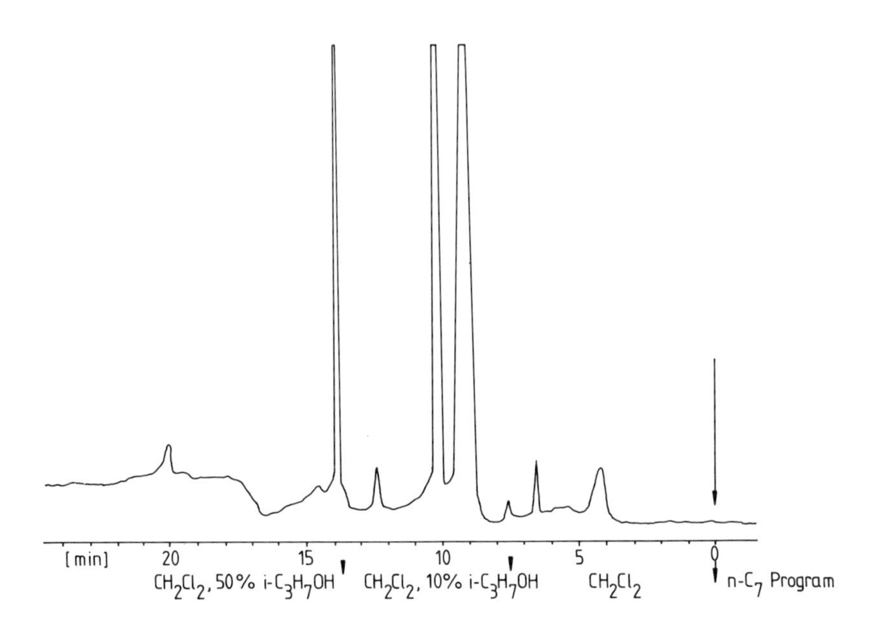

Fig.VI.21. Multi-step gradient elution. Blank gradient. Same conditions as in Fig.VI.20, but without sample and initial isocratic period. (Eluent impurities!)

Fig.VI.20. It is surprising how many and in what concentrations the impurities are collected by the column. Therefore, the regeneration times or flush volumes should be kept constant between or before each analysis. It is advisable that a blank gradient elution be run before or after a separation.

These enrichment effects occur with both polar and nonpolar stationary phases. Not only are the impurities in the first (weaker) gradient component collected on the column, but also those contained in the stronger solvent which are not displaced because of the low initial elution strength of the gradient mixture. This means that only the purest solvents (reagent grade or those for "residue analysis") can be used for gradient elution - an expensive proposition since the eluents are mixtures at the end and can hardly be recovered.

Column regeneration, i.e., the return to the original conditions by removing the polar eluent component, requires a long time, especially with polar stationary phases. Difficulties stem less from flushing out the residual polar eluent than from re-establishing equilibrium between the water on the adsorbent and in the eluent [69]. The reinstitution of the starting conditions should be verified by checking the k' values of test substances. At least 10 to 30 column volumes are necessary for the regeneration. Since this frequently requires more time than the gradient-elution separation itself, it is often more practical not to return to the original conditions (equilibrium with nonpolar eluents), but to set a standard regeneration time. The latter is indispensable anyway because of the eluent impurities. The regeneration time can also be shortened considerably by reversing the gradient [69,70].

Nonpolar and chemically bonded phases are regenerated with fewer difficulties. Equilibrium is re-established after passing through at least 10 column volumes of eluent. However, the regeneration time should be standardized in these cases also, as the impurities in water (or) methanol are enriched on nonpolar phases as well. To save time, the regeneration may, of course, be performed at higher flow rates than the separations. The regeneration of columns after incremental gradient elution, i.e., returning from water to heptane as eluent, requires a sequence of three or five solvents in order to avoid demixing effects [65].

Gradient elution is essential for the separation of mixtures having a wide range of polarities. Such problems appear in adsorption chromatography with polar and nonpolar stationary phases and in ion-exchange separations (cf. VIII.D.5). Adsorption chromatography can be readily used to separate mixtures with a k' range of 10^4 or more in a single analysis. In routine applications these tremendous advantages are offset by a few difficulties whose causes were explained in order to avoid possible misinterpretations of gradient-elution chromatograms. By observing the precautionary measures described, such as carrying out a blank gradient elution run preceding or following an analysis, gradient elution can be used without further ado.

If the expensive equipment for HPLC gradient elution (mixing on the high pressure side) is unvailable, gradient mixing may be performed before the pump (on the low pressure side) [63]. Alternatively, the solvent may be changed stepwise by simply changing the reservoir. By making these steps sufficiently small, the result corresponds completely to gradient elution. A chromatogram obtained in this way is shown in Fig.VI.22 [71] for the separation of the phenacyl esters of fatty acids. Due to the low flow rate used (0.1 cm/sec), the separation required more than 4 hours.

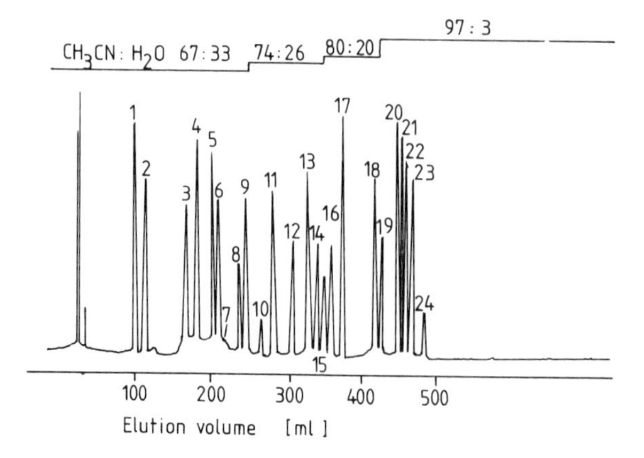

Fig.VI.22. Stepwise elution. Separation of the phenacyl esters of fatty acids [71]. Column: 90 cm, 6.4 mm i.d. Stationary phase: µ-Bondapak C_{18} (Waters Asso.). Eluent: acetonitrile-water mixtures; F = 2 cm^3/min; u ~ 1 mm/sec. Samples: saturated and unsaturated fatty acids from C_{12} (lauric acid) to lignoceric acid (C_{24})

IV. Applications of Adsorption Chromatography

In recent years the number of HPLC applications has grown exponentially. The first monographs and review articles covering HPLC application in clinical chemistry [72], the pharmaceutical industry [73], the sequencing of peptides and proteins [74], and in forensic chemistry [75] have already appeared. It is beyond the scope of this introductory text to present all the separations that have been published since then. Their inclusion in the individual chapters on the various separation systems would have been largely arbitrary. Here, reference is made to the monographs and bibliographies that discuss the separations which are arranged by class of substances [76-78]. In this respect, the older monographs on column chromatography [79,80] are very useful because the classical column chromatographic systems can be adapted quite readily to HPLC. Here only certain selected examples of separations are presented that demonstrate the application potentials of the various separation systems.

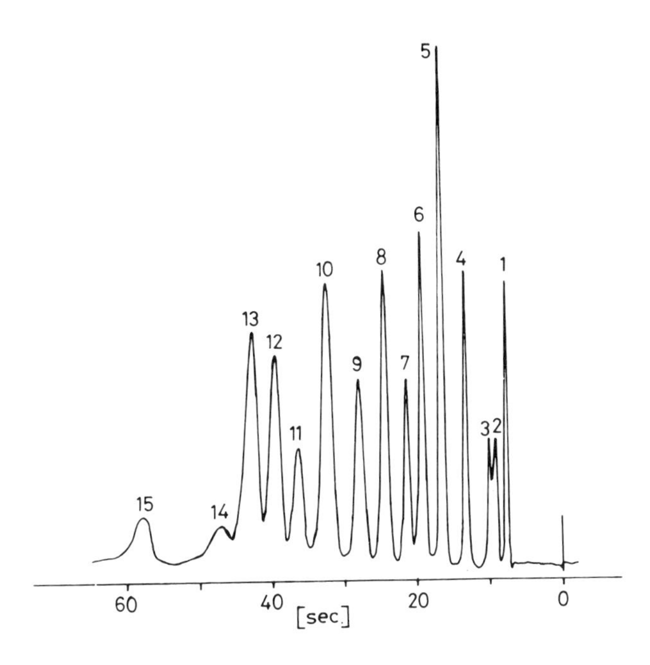

Fig.VI.23. Separation of condensed aromatics. Silica, Lichrospher (Merck); d_p ~ 4.5 µm; eluent: n-heptane; column: 6.5 cm, 4 mm i.d.; Δp = 72 atm; u = 9.3 mm/sec. 1 = unretained; 2 = solvent; 3 = benzene; 4 = naphthalene; 5 = diphenyl; 6 = anthracene; 7 = pyrene; 8 = fluoranthene; 9 = o-terphenyl; 10 = 1,2-benzanthracene; 11 = 3,4 benzpyrene; 12 = perylene; 13 = 1,12-benzperylene; 14 = coronene; 15 = 1,2,5,6-dibenzanthracene

A. On Polar Stationary Phases

The rules established for classical column chromatography (adsorption chromatography) are also valid for HPLC on polar solids. Hence the primary area of application deals with the separation of nonpolar to moderately polar organic compounds. Strongly polar and ionic compounds are retained too strongly and must therefore be separated by means of other systems (partition or ion-exchange chromatography). Nonpolar stationary phases provide considerable advantages for such separations.

Polycyclic aromatic hydrocarbons have been separated on silica, alumina, porous solids, and PLB (cf. Fig.VI.3 and 6) [13,81-85]. To enhance the selectivity, the adsorbents have been coated with complexing agents (silver nitrate, trinitrofluorenone) [3,4]. The cis-trans isomers of olefinic acetates (C_{10} to C_{18}) were separated on $AgNO_3$-coated silica [86], or the complexing agent ($AgNO_3$) was added to the eluent [87]. The efficiency of HPLC in this field is illustrated by Fig.VI.23 via the separation of 13 aromatic hydrocarbons in 60 sec. This separation demonstrates furthermore that relatively short columns (6.5 cm) containing very small particles (dp ~ 5 μm) generate sufficient theoretical plates (N ~ 3000) to perform even such difficult separations. Because of the shortness of the column, the required pressure drop was surprisingly small (Δp = 72 atm) despite the small particle size used.

Pesticides [62,88], polychlorinated biphenyls [89,90], and herbicides [91] have also been separated (cf. Fig.VI.13). However, the detection sensitivity of selective GC detectors could not be achieved [92]. Nevertheless, such substances can be separated by LC without difficulties from decomposition and without derivatization (hence significantly simpler).

From the beginning the principal application of HPLC has been in the analysis of pharmaceuticals and pharmaceutical preparations [93-95], steroids [96,96a], cardiac glycosides [97,98], alkaloids [99], and aflatoxins [100]. Diastereomers have also been separated [101,102]. Thus, Nakanishi and co-workers [103] used a recycling system to increase the separation efficiency. They used silica with a large particle size (40 μm) and pumped the mixture several times through the same column. For 4 - 8 μm silica even a 30 cm column provides enough efficiency to resolve the pair of diastereomers [102,104, 105] (Fig.VI.24). Porphyrin and chlorophyl derivatives (methyl esters) can also be separated on polar stationary phases [106,107]. But even very polar substances such as nonionic emulsifiers based on polyethylenglycol could be separated via gradient elution [147] on silica.

It should be pointed out that many separations can be carried out on other stationary phases (e.g., reversed phases) and systems (e.g., partition chromatography). The choice of the system depends not only on the compounds to be separated, but also on the experiences accrued with a particular separation system.

B. On Nonpolar Stationary Phases

Today, separation on reversed phases is most frequently employed in HPLC. It is estimated that 60 - 80% of all analytical separations

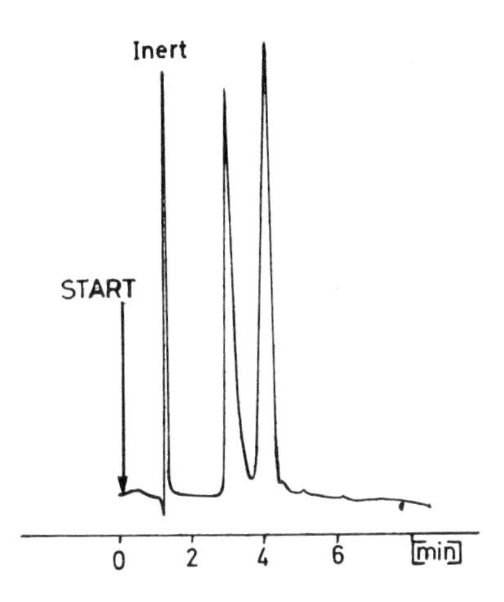

Fig.VI.24. Separation of Diastereomers (O-methylmandelic acid(-)α-phenetylamide [104]. Silica, Spherosil XOA 400; eluent: isooctane-chloroform (2:1); column: 30 cm, 3 mm i.d.; Δp = 135 atm; F = 1.4 ml/min

are performed by this technique. In most cases silica gels reacted with octadecylsilanes are used as stationary phase and mixtures of water with methanol or acetonitrile as eluents. Such systems are very simple and have good reproducibility. These eluents can be readily prepared in sufficient purity. The columns are stable for longer periods at pH < 8.

The areas of application of polar and nonpolar stationary phases overlap. A comparison of separations conducted on a polar stationary phase (a chemically bonded ether) using a nonpolar eluent and on the same phase with a polar eluent (by the RP method) is presented in Fig.VI.25 [108]. As expected, the elution order of substituted ureas is reversed, so that the compound with the longest alkyl group (Neburan) is retained most strongly on the RP system.

Even aromatic hydrocarbons and pesticides [109-111] have been separated on reversed phases. Benzene could be separated from deuterobenzene with a methanol-water eluent [112]. Since nonpolar substances are retained very strongly on RP from water, such phases are suitable for the enrichment of trace organics from water [113,114] or marine sediments [115].

Corticosteroids (Fig.VI.26), including androgenic hormones and progesterones [38], phythosterines [116], cardiac glycosides [38,117], bufotoxins [118] can be readily separated by various water-methanol

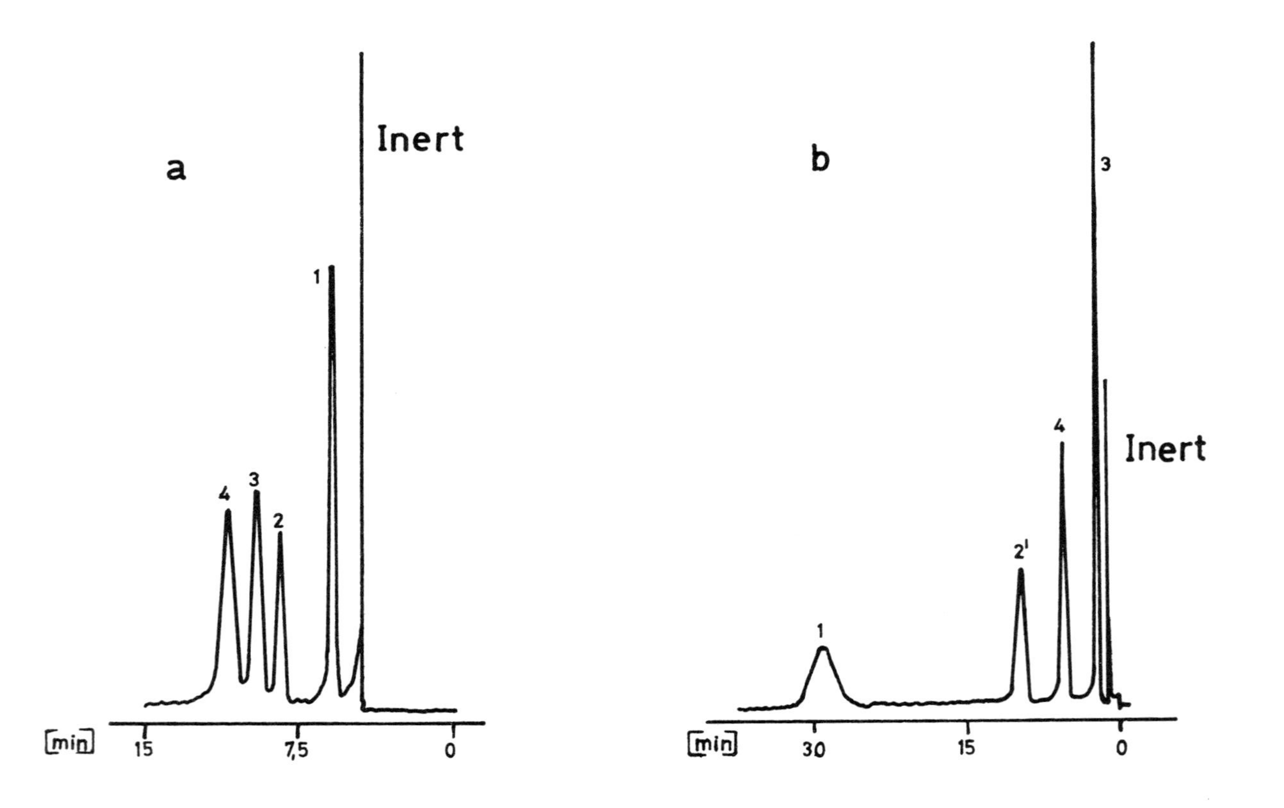

Fig.VI.25. Elution sequence for normal and reversed phase chromatography [108]. Separation of substituted ureas. Chemically bonded phase, Permaphase-ETH; $d_p \sim 27$ µm. Eluent: a) hexane + 1% dioxane, b) water-methanol (65:35). Column: 100 cm, 2 mm i.d. a) $\Delta p = 25$ atm; F = 1 ml/min; T = 27°C. b) $\Delta p = 60$ atm; F = 1 ml/min; T = 50°C. 1 = Neburon; 2 = Fenuron; 3 = Linuron; 4 = Diuron (methyl + phenyl-substituted ureas)

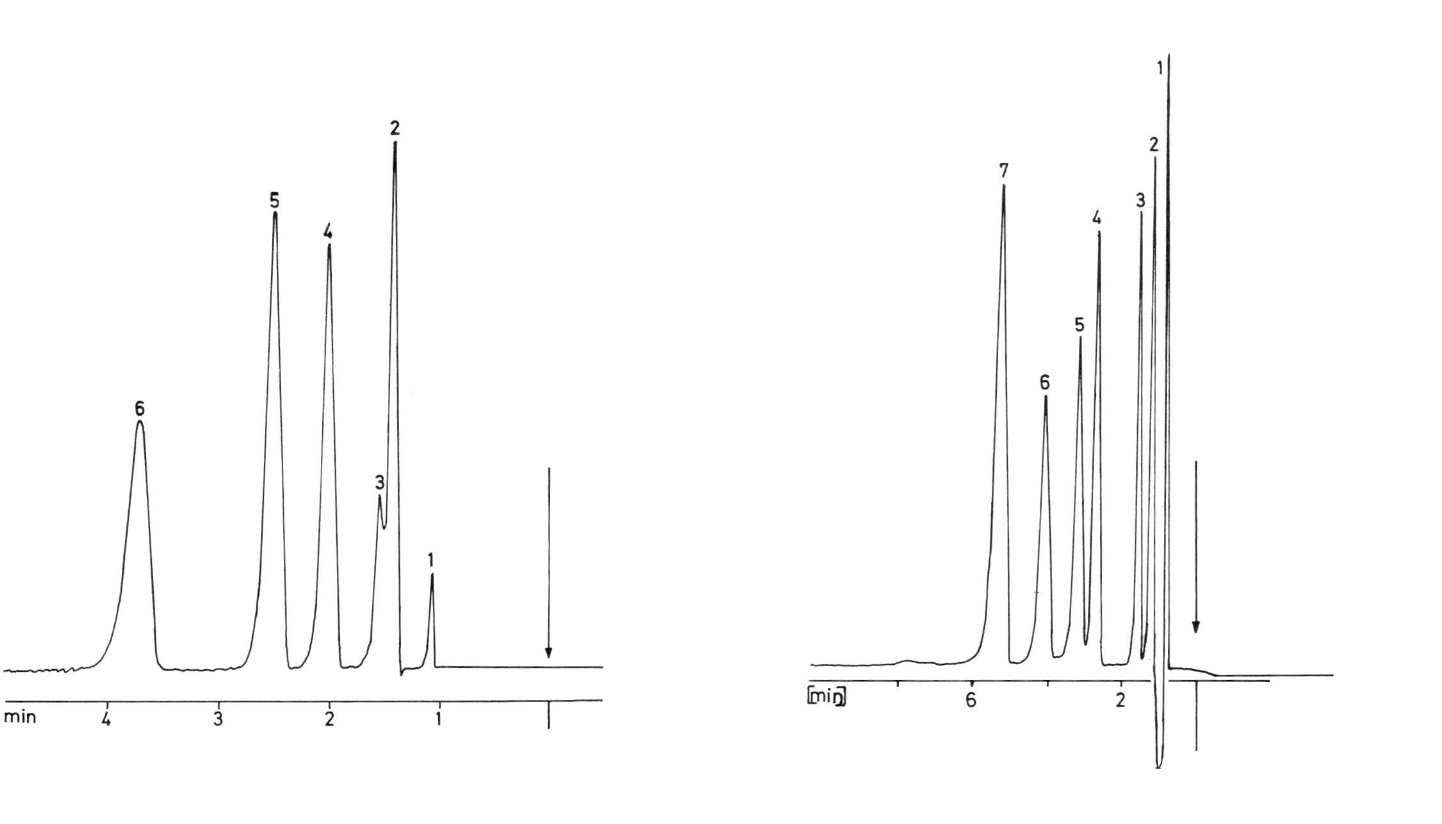

Fig.VI.26. Separation of corticosteroids. Reversed phase: octadecylsilane on Si 100; d_p ~ 10 µm. Eluent: water-methanol (25:75). Column: 30 cm, 4 mm i.d.; u = 5.2 mm/sec; Δp = 175 atm. 1 = unretained; 2 = cortisone; 4 = tetra-hydrocortisone; 5 = 11-desoxycorticosterone; 6 = 11-desoxycorticosterone acetate

Fig.VI.27. Separation of phenolcarboxylic acids. Reversed phase: octadecylsilane on Si 100; d_p ~ 10 µm. Eluent: water + acetic acid (80:20). Column: 30 cm, 4 mm i.d.; u = 5.9 mm/sec; Δp = 160 atm. 1 = China acid + methanol; 2 = chlorogenic acid; 3 = caffeic acid; 4 = hydroxycinnamic acid; 5 = 3-hydroxycinnamic acid; 6 = 2-hydroxycinnamic acid; 7 = coumarin

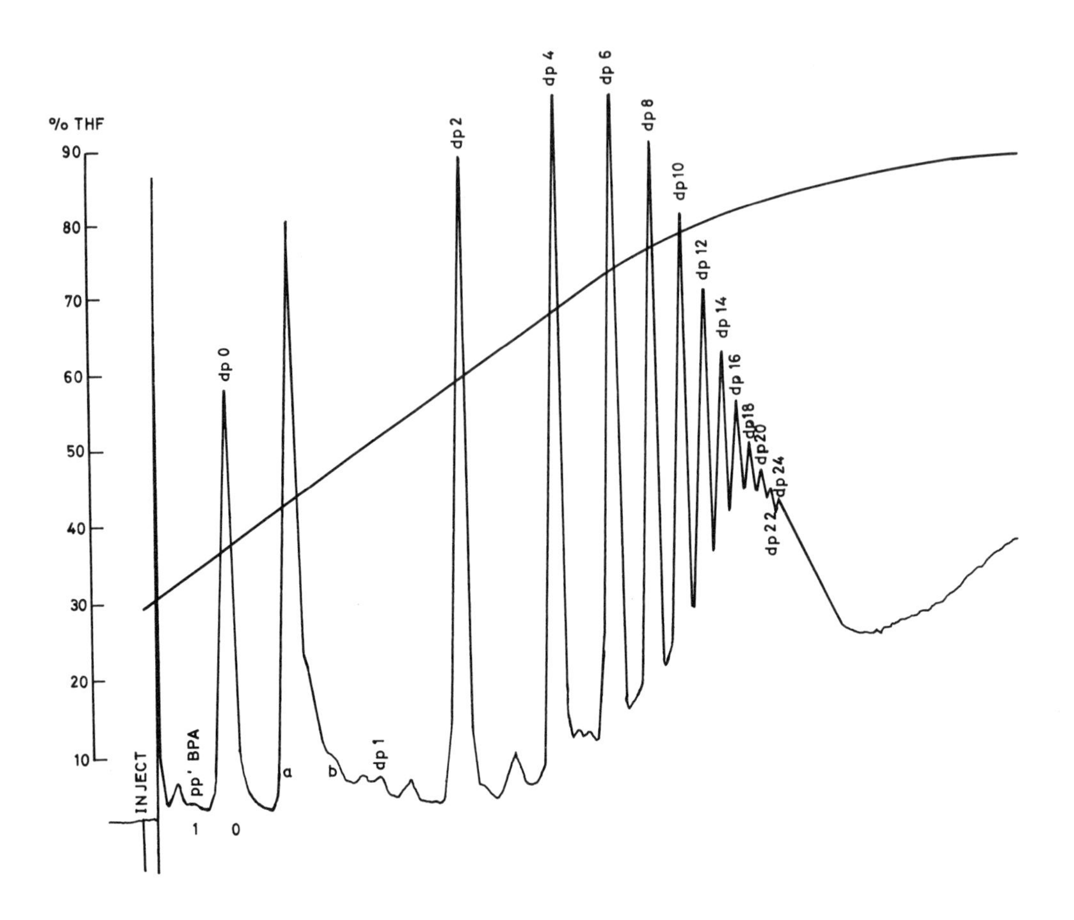

Fig.VI.28. Characterization of an epoxy resin - Epon 1004 [133]. Column: 120 cm, 2 mm i.d.; stationary phase: Corasil-Bondapak C_{18}; d_p ~ 40 µm; eluent: gradient from 30% tetrahydrofuran in water to 90% tetrahydrofuran in water. F = 1 cm^3 /min

mixtures, or detected in biological materials [119,120,138-141]. Carboxylic acids [38] (Fig.VI.27) or their derivatives (Fig.VI.22) [71, 121], amino acids [122] or their phenylhydantoin derivatives [123-125], peptides [122,126,137], antibiotics [127,128], prostaglandins [129], and other polar substances such as ergotamine derivatives, LSD [130], N-nitroso compounds [131], and fluorescamine derivatives [132] have been separated on reversed phases and detected in pharmaceutical preparations or body fluids.

On reversed phases sorption increases with the number of nonpolar groups. Homologous series of organic compounds, as well as members of a homologous polymer series, are readily separated on RP. Using a methanol-methylene chloride eluent, polystyrenes up to a molecular weight of *ca.* 3000 could be resolved into the individual members and separated on a preparative scale [43].

Fig.VI.28 shows the separation of an epoxy resin on an RP [133] where gradient elution with an increasing tetrahydrofuran content was used to reduce the analysis time.

Even polar, highly water-soluble compounds such as pentaerythritol [134] and the oligosaccharides can be separated on RP, whereas mono- and disaccharides can be resolved on a chemically bonded phase with an amino group [135] (e.g., chemically bonded 3-aminopropylsilane), as shown in Fig.VI.29 [133].

The bonding of organic molecules other than alkylsilanes to the silica surface always diminishes its adsorption activity. Such stationary phases are suitable even for the separation of polar substances using eluent mixtures of moderate polarity. In addition, their selectivity is altered, and their properties depend strongly on the extent of coverage by the organic moieties. Stationary phases are difficult to prepare with different functional groups but having the same extent of coverage. Hence, the exact effect of functional groups is difficult to ascertain. In addition to the amino and carbohydrate phases already mentioned, those containing nitro, nitrile, and glycol groups are also commercially available. The latter appear to have utility for the aqueous exclusion chromatography of peptides (cf. Chapter IX). With polar eluents such as methanol-water the RP properties seem to predominate, whereas with the nonpolar the polar functional groups and the unshielded silanol groups of the support are both responsible for the selectivity.

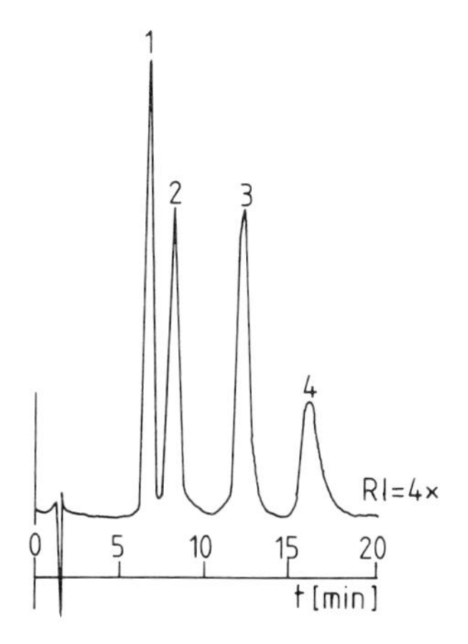

Fig.VI.29. Separation of sugars [133]. Column: 30 cm, 4 mm i.d.; stationary phase: μ-Bondapak-Carbohydrate; d_p ~ 10 μm; eluent: acetonitrile-water (75:25). 1 = Fructose; 2 = glucose; 3 = sucrose; 4 = maltose

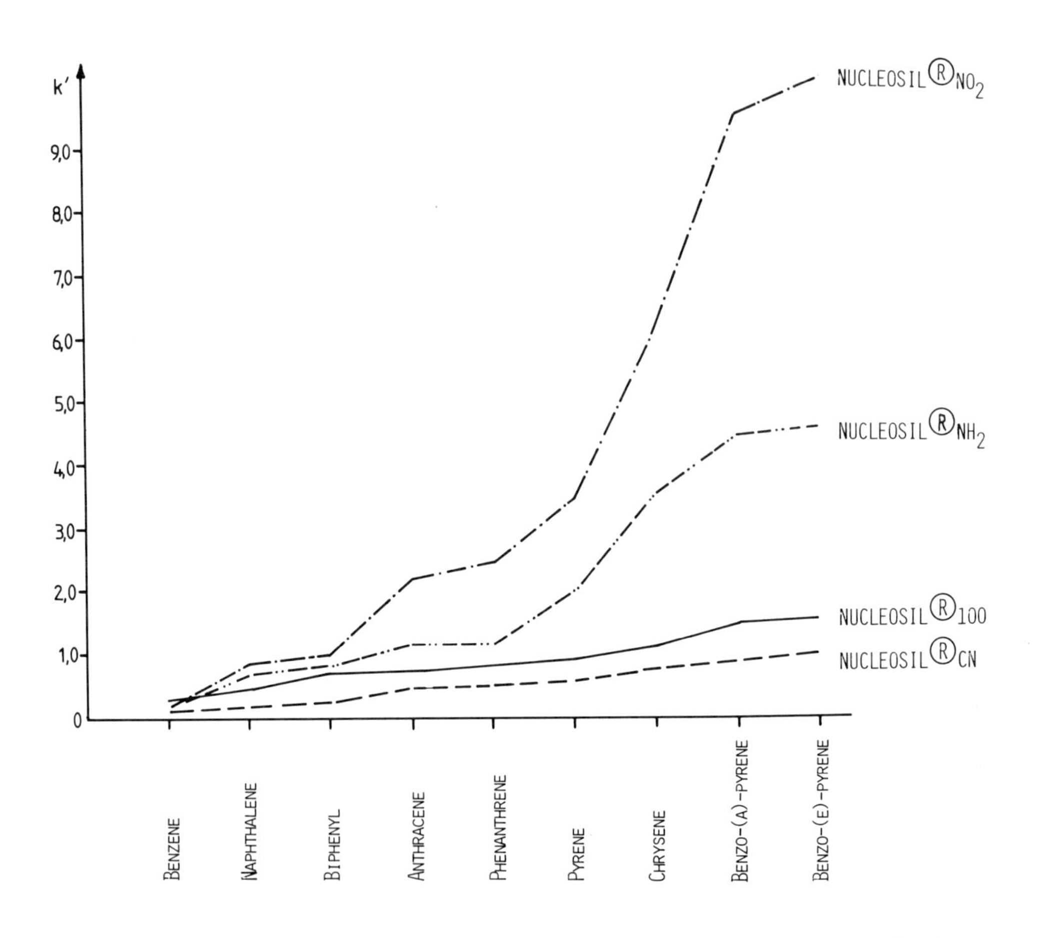

Fig.VI.30. Selectivity of bonded phases with different functional groups [149]

Fig.VI.30 compares the selectivity of chemically bonded phases containing various functional groups with that of the original silica gel. On the phase containing nitro groups the k' values for the condensed aromatics were about 5 times those observed on silica gel, using water-saturated heptane as eluent. On the nitrile phase all k' values were less than 1. The bonded amino phase (which is suitable for the separation of sugars in aqueous solution) also exhibited better selectivity in resolving aromatics than the original silica.

Differences in selectivity between an ordinary RP and a nitrile phase have been observed [136] in the separation of alkylphenols. Others have reported that the selectivity of the nitrile phase obtained from one company was similar to that of the amino phase from another, whereas the nitrile phases from both were not comparable in their selectivity [148]. Hence, it appears that the standardization of chemically bonded phases may pose some of the same problems encountered with liquid phases in gas chromatography.

C. Separations on Polyamide

The polarity of the commercial polyamide phases is between that of the silica gel and the RP systems. However, the rate of mass transfer on them is slower than on silica gel or the brush-type stationary phases because of the slow diffusion in the polymer stationary phase. This type of support material represents a transition between chemically modified phases and partition chromatographic systems. Quinones, phenols, sulfonamides, food preservatives [142-144], dansyl-amino acids [145], and metalloporphyrins [146] can be separated on polyamides.

References Chapter VI

1. Snyder, L.R.: Principles of Adsorption Chromatography. New York, N.Y.: Marcel Dekker 1968
2. Böhme, W.: Dissertation, Universität Saarbrücken 1976
3. Vivilecchia, R., Thiebaud, M., Frei, R.W.: J. Chromatogr. Sci. *10*, 411 (1972)
4. Karger, B.L., Guiochon, G., Martin, M., Lohéac, J.: Anal. Chem. *45*, 496 (1973)
5. Brust, E.O., Sebestian, I., Halász, I.: J. Chromatogr. *83*, 15 (1973)
6. Locke, D.C.: J. Chromatogr. Sci. *11*, 120 (1973)
7. Brockmann, H., Schodder, H.: Ber. dtsch. chem. Ges. *74*, 73 (1941)
8. Lederer, E. (Ed.): Chromatographie en chimie organique et biologique. Paris: Masson 1959
9. Hesse, G.: Chromatographisches Praktikum. Frankfurt/Main: Akadem. Verlagsgesellschaft 1968
10. Hesse, G., Roscher, G.: Z. anal. Chem. *200*, 3 (1965)
11. Dr.Eisenbeiß, Merck AG., Darmstadt: Courtesy of
12. Engelhardt, H., in M. Zief, (Ed.): Ultrapurity. New York: Dekker 1972
13. Engelhardt, H., Böhme, W.: J. Chromatogr. *133*, 67 (1977)
14. Snyder, L.R.: J. Chromatogr. Sci. *7*, 595 (1969)

15. Snyder, L.R., Kirkland, J.J.: An Introduction to Modern Liquid Chromatography. New York: Wiley-Interscience 1974
16. Wiedemann, H.: Diplomarbeit Erlangen 1971, vgl.: Engelhardt, H., Wiedemann,H.: Anal. Chem. *45*, 1649 (1973)
17. Maggs, R.J.: J. Chromatogr. Sci. *7*, 145 (1969)
18. Scott, R.P.W., Lawrence, J.G.: J. Chromatogr. Sci. *8*, 619 (1970)
19. Kirkland, J.J.: J. Chromatogr. *83*, 149 (1973)
20. Snyder, L.R.: J. Chromatogr. *63*, 15 (1971)
21. Snyder, L.R.: Anal. Chem. *46*, 1384 (1974)
22. Engelhardt, H.: J. Chromatogr. Sci. *15*, 380 (1977)
23. Trappe, W.: Biochem. Z. *305*, 150 (1940)
24. Neher, R., in: "Thin Layer Chromatography", G.B. Marini-Bettolo Ed. Amsterdam: Elsevier 1964
25. Saunders, D.L.: Anal. Chem. *46*, 470 (1974)
26. Snyder, L.R.: Anal. Chem. *46*, 1384 (1974)
27. Geiss, F.: Die Parameter der Dünnschicht-Chromatographie. Braunschweig: Vieweg 1972
28. Howard, G.A., Martin, A.J.P.: Biochem. J. *46*, 215 (1951)
29. Majors, R.E.: Analysis *10*, 549 (1975)
30. Collin, H., Eon, C., Guiochon, G.: J. Chromatogr. *119*, 41 (1976)
31. Collin, H., Eon, C., Guiochon, G.: J. Chromatogr. *122*, 223 (1976)
32. Collin, H., Guiochon, G.: J. Chromatogr. *126*, 43 (1976)
33. Collin, H., Guiochon, G.: J. Chromatogr. *141*, 289 (1977)
34. Karch, K., Sebestian, I., Halâsz, I.: J. Chromatogr. *122*, 3 (1976)
35. Kirkland, J.J.: Chromatographia *8*, 661 (1975)
36. Hemetsberger, H., Maasfeld, W., Ricken, H.: Chromatographia *9*, 303 (1976)
37. Landolt-Börnstein, 6. Auflage, 2. Band, 5. Teil, Bandteil a Transportphänomene I
38. Karch, K., Sebestian, I., Halâsz, I., Engelhardt, H.: J. Chromatogr. *122*, 171 (1976)
39. Karger, B.L., Gant, J.R., Hartkopf, A., Weiner, P.H.: J. Chromatogr. *128*, 65 (1976)
40. Horvath, C., Melander, W., Molnâr, I.: J. Chromatogr. *125*, 129 (1976)
41. Horvath, C., Melander, W., Molnâr, I.: Anal. Chem. *49*, 142 (1977)
42. Helfferich, F.H.: Ionenaustauscher. Weinheim: Verlag Chemie 1959
43. Werner, W.: Dissertation Saarbrücken 1976
44. Schneider, W., Bruderreck, H., Halâsz, I.: Anal. Chem. *36*, 1533 (1964)
45. Roth, B.: Dissertation Saarbrücken 1977
46. Vonach, B., Schomburg, G.: J. Chromatogr. *149*, 417 (1978)
47. Snyder, L.R.: J. Chromatogr. Sci. *8*, 692 (1970)
48. Wiedemann, H., Engelhardt, H., Halâsz, I.: J. Chromatogr. *91*, 141 (1974)
49. Halâsz, I., Holdinghausen, F.: J. Gas Chromatogr. *5*, 385 (1967)
50. Scott, R.P.W., Lawrence, J.G.: J. Chromatogr. Sci. *7*, 65 (1969)

51. Maggs, R.J.: J. Chromatogr. Sci. *7*, 145 (1969)

52. Wiedemann, H.: Dissertation Saarbrücken 1973

53. Kikta jr., E.J., Stange, H.E., Lam, S.: J. Chromatogr. *138*, 321 (1977)

54. Hesse, G., Roscher, G.: Chromatographia *2*, 512 (1969)

55. Berry, L.V., Engelhardt, H.: J. Chromatogr. *95*, 27 (1974)

56. Snyder, L.R.: J. Chromatogr. *13*, 415 (1964)

57. Simmons, M.C., Snyder, L.R.: Anal. Chem. *30*, 32 (1958)

58. Szonntagh, E.L., in: Ettre, L.S., Zlatkis, A. (Ed.): The Practice of Gas Chromatography. New York: Interscience 1967

59. Spackman, D.H., Stein, W.H., Moore, S.: Anal. Chem. *30*, 1190 (1958)

60. Scott, C.D., Chilcote, D.D., Lee, N.A.: Anal. Chem. *45*, 85 (1972)

61. Huber, J.F.K., van der Linden, R., Ecker, E., Oreans, M.: J. Chromatogr. *93*, 267 (1973)

62. Dolphin, R.J., Willmot, F.W., Mills, A.D., Hoogeveen, L.P.J.: J. Chromatogr. *122*, 259 (1976)

63. Scott, R.P.W., Kucera, P.: Anal. Chem. *45*, 749 (1973)

64. Scott, R.P.W., Kucera, P.: J. Chromatogr. Sci. *11*, 83 (1973)

65. Scott, R.P.W., Kucera, P.: J. Chromatogr. *83*, 257 (1973)

66. Snyder, L.R., in: Kirkland, J.J.: Modern Practice of Liquid Chromatography, p. 220 ff. New York: Wiley-Interscience 1971

67. Snyder, L.R., Saunders, D.L.: J. Chromatogr. Sci. *7*, 195 (1969)

68. Elgass, H.: Ph.D.-Thesis, University Saarbrücken 1978

69. Engelhardt, H., Elgass, H.: J. Chromatogr. *112*, 415 (1975)

70. Majors, R.E.: Anal. Chem. *45*, 755 (1973)

71. Borch, R.F.: Anal. Chem. *45*, 755 (1973)

72. Dixon, P.F., Gray, C.H., Lim, C.K., Stoll, L.S.: High Pressure Liquid Chromatography in Clinical Chemistry. London - New York - San Francisco: Academic Press 1976

73. Bailey, F.: J. Chromatogr. *122*, 73 (1976)

74. Deyl, Z.: J. Chromatogr. *127*, 91 (1976)

75. Wheals, B.B.: J. Chromatogr. *122*, 85 (1976)

76. Parris, N.A.: Instrumental Liquid Chromatography. Amsterdam: Elsevier 1976

77. Deyl, Z., Macek, K., Janak, J. (Ed.): Liquid Column Chromatography. Amsterdam: Elsevier 1975

78. Deyl, Z., Kopecky, J.: Bibliography of Liquid Column Chromatography. J. Chromatogr., Suppl. Vol. 6, 1976

79. Heftmann, E.: Chromatography 3. Ed. New York: Van Nostrand-Reinhold 1975

80. Lederer, E.: Chromatographie en Chimie organique et biologique. Vol. I et Vol. II. Paris: Masson 1959/1960

81. Strubert, W.: Chromatographia *6*, 205 (1973)

82. Martin, M., Lohéac, J., Guiochon, G.: Chromatographia *5*, 33 (1972)

83. Popl, M., Stejskal, M., Mostecký, J.: Anal. Chem. *46*, 1581 (1974)

84. Boden, H.: J. Chromatogr. Sci. *14*, 391 (1976)

85. Böhme, H., Engelhardt, H.: Compendium 74/75, Ergänzungsband zu "Erdöl und Kohle" 1975

86. Heath, R.R., Tumlinson, J.H., Doolittle, R.E., Proveaux, A.T.: J. Chromatogr. Sci. *13*, 380 (1975)

87. Schomburg, G., Zegarski, K.: J. Chromatogr. *114*, 174 (1975)

88. Little, J.N., Horgan, D.F., Bombaugh, K.J.: J. Chromatogr. Sci. *8*, 625 (1970)

89. Brinkman, K.A.Th., Seetz, J.W.F.L., Reymer, H.G.M.: J. Chromatogr. *116*, 353 (1976)

90. Brinkman, K.A.Th., De Kok, A., De Vries, G., Reymer, H.G.M.: J. Chromatogr. *128*, 101 (1976)

91. Eisenbeis, F., Sieper, H.: J. Chromatogr. *83*, 439 (1973)

92. Jork, H., Roth, B.: J. Chromatogr. *144*, 39 (1977)

93. Hinsvark, O.N., Zazulak, W., Cohen, A.I.: J. Chromatogr. Sci. *10*, 379 (1972)

94. Krol, G.J., Mannan, C.A., Gemmill jr., F.Q., Hicks, G.E., Uko, B.T.: J. Chromatogr. *74*, 43 (1972)

95. Stutz, M.H., Sass, S.: Anal. Chem. *45*, 2134 (1973)

96. Fitzpatrick, F.A., Siggia, S., Dingman, J.: Anal. Chem. *44*, 2211 (1972)

96a. Touchstone, J.C., Wortmann, W.: J. Chromatogr. *76*, 244 (1973)

97. Castle, M.C.: J. Chromatogr. *115*, 437 (1975)

98. Nachtmann, F., Spitzy, R.W., Frei, R.W.: J. Chromatogr. *122*, 293 (1976)

99. Erni, F., Frei, R.W., Lindner, W.: J. Chromatogr. *125*, 265 (1976)

100. Seitz, L.M.: J. Chromatogr. *104*, 81 (1975)

101. Helmchen, G., Haas, G., Prelog, V.: Helv. Chim. Acta *56*, 2255 (1973)

102. Helmchen, G., Strubert, H.: Chromatographia *7*, 713 (1974)

103. Koreeda, M., Weiss, G., Nakanishi, K.: J. Am. Chem. Soc. *95*, 239 (1973)

104. Siemens, Karlsruhe: Anwendungsbeispiel 12/03

105. Scott, C.G., Petrin, M.J., McCorkle, T.: J. Chromatogr. *125*, 157 (1976)

106. Evans, N., Games, D.E., Jackson, A.H., Matlin, S.A.: J. Chromatogr. *115*, 325 (1975)

107. Evans, N., Jackson, A.H., Matlin, S.A., Towill, R.: J. Chromatogr. *125*, 345 (1976)

108. Kirkland, J.J.: Anal. Chem. *43*, (12) 43 a (1971)

109. Vaughan, C.G., Wheals, B.B., Whitehouse, M.J.: J. Chromatogr. *78*, 203 (1973)

110. Seiber, J.N.: J. Chromatogr. *94*, 151 (1974)

111. Klimisch, H.J., Ambrosius, D.: J. Chromatogr. *120*, 299 (1976)

112. Cartoni, G.P., Ferretti, J.: J. Chromatogr. *122*, 287 (1976)

113. Aufsatz, M.: Diplomarbeit, Saarbrücken 1976

114. Leoni, V., Puccetti, G., Grella A.: J. Chromatogr. *106*, 119 (1975)

115. May, W.E., Chesler, S.N., Cram, S.P., Gump, B.H., Hertz, H.S., Enagonio, O.P., Dyszel, S.M.: J. Chromatogr. Sci. *13*, 535 (1975)

116. Rees, H.H., Donnahey, D.L., Goodwin, T.W.: J. Chromatogr. *116*, 281 (1976)

117. Erni, F., Frei, R.W.: J. Chromatogr. *130*, 169 (1977)

118. Shimada, K., Hasegawa, M., Hasebe, K., Fujii, Y., Nambara, T.: J. Chromatogr. *124*, 79 (1976)

119. O'Hare, M.J., Nice, E.C., Magee-Brown, R., Bullman, H.: J. Chromatogr. *125*, 357 (1976)

120. Benson, J.M., Seiber, J.N.: J. Chromatogr. *148*, 521 (1978)

121. Hoffmann, N.E., Liao, J.C.: Anal. Chem. *48*, 1104 (1976)

122. Molnar, J., Horvath, C.: J. Chromatogr. *142*, 623 (1977)

123. Graffeo, A.P., Haag, A., Karger, B.L.: Analyt. Letters *6*, 305 (1973)

124. Haag, A., Langer, K.: Chromatographia *7*, 659 (1974)

125. Margolies, M.N., Brauer, A.: J. Chromatogr. *148*, 429 (1978)

126. Krummen, K., Frei, R.W.: J. Chromatogr. *132*, 27, 429 (1977)

127. Chevalier, G., Bollet, C., Rohrbach, P., Risse, C., Chaude, M., Rosset, R.: J. Chromatogr. *124*, 343 (1976)

128. Hartmann, V., Rödiger, M.: Chromatographia *9*, 266 (1976)

129. Fitzpatrick, F.A.: Anal. Chem. *48*, 499 (1976)

130. Jane, I., Wheals, B.B.: J. Chromatogr. *84*, 181 (1973)

131. Heyns, K., Röper, H.: J. Chromatogr. *93*, 429 (1974)

132. Samejima, K.: J. Chromatogr. *96*, 250 (1974)

133. Application manuals by Waters Assoc., Milford, Mass., U.S.A.

134. Callmer, K.: J. Chromatogr. *115*, 397 (1975)

135. Jones, A.D., Burns, J.W., Sellings, S.G., Cox, J.A.: J. Chromatogr. *144*, 169 (1977)

136. Callmer, K., Edholm, L.E., Smith, B.E.F.: J. Chromatogr. *136*, 45 (1977)

137. Hansen, J.J., Greibrokk, T., Currie, B.L., Johansson, K.N.-G., Folkers, K.: J. Chromatogr. *135*, 155 (1977)

138. Dolphin, R.J., Pergande, P.J.: J. Chromatogr. *143*, 267 (1977)

139. Loo, J.C.K., Butterfield, A.G., Moffat, J., Jordan, N.: J. Chromatogr. *143*, 275 (1977)

140. Loo, J.C.K., Jordan, N.: J. Chromatogr. *143*, 314 (1977)

141. Schwedt, G., Bussemas, H.H., Lippman, Ch.: J. Chromatogr. *143*, 259 (1977)

142. Rabel, F.M.: Anal. Chem. *45*, 957 (1973)

143. Olson, L., Samuelson, O.: J. Chromatogr. *106*, 139 (1975)

144. Collet, G., Rocca, J.L., Sage, D., Berticat, P.: J. Chromatogr. *121*, 213 (1976)

145. Deyl, Z., Rosmus, J.: J. Chromatogr. *69*, 129 (1972)

146. Svrivastava, T.S., Yonetani, T.: Chromatographia *8*, 124 (1975)

147. Brüschweiler, H.: Mitt. Gebiete Lebensm. Hyg. *68*, 46 (1977)

148. Qualls jr., C.W., Segall, H.J.: J. Chromatogr. *150*, 202 (1978)

149. Sebestian, I., Macherey & Nagel, Düren, BRD: Courtesy of

Chapter VII

Partition Chromatography

A. Introduction

Separations are based on differences in the distribution of the sample components between the stationary phase (the liquid phase coated on the support) and the mobile phase that flows through the packing. The support for the stationary phase should not affect the partition equilibrium (i.e., it should be *inert*), but should only provide uniform distribution of the stationary liquid over a large surface area so that rapid equilibration of the sample between the two phases is achieved.

The underlying principle of this method is the partitioning (or distribution) of substances between two *liquid* phases. It is similar to the extractions used in basic laboratory operations. The repeated (or multiple) application of such partition steps for the separation of mixtures [1] had already been used for a long time before Martin and Synge [2] adapted it to the usual chromatographic method by coating one of the liquid phases onto a porous, finely-divided powder (the support). This dry-looking powder (the stationary phase) was packed in a column and the second liquid (the mobile phase) was allowed to percolate through the packing. If the support is inert, the distribution coefficients determined by the classical method will agree with those obtained from a chromatogram.

Partition processes have found extensive application in gas chromatography where almost all separations are based on the distribution between the liquid stationary phase and the gaseous mobile phase. The partition processes are theoretically well understood and can be carried out reproducibly because of the well characterized properties of the liquid phase. Therefore, there are many papers on the theoretical treatment of partition processes which calculate physico-chemical quantities from gas [3,4] and liquid chromatographic [5] data.

Some difficulties arising from the nature of the system itself prevent the general application of the partition process in liquid chromatography. The two liquid phases should *not* be miscible. Since this is seldom the case, both liquids must be mutually equilibrated prior to preparation of the stationary phase (support plus liquid phase). Even with good saturation of the mobile phase with the liquid phase, a precolumn packed with support heavily loaded with the liquid phase is necessary to fully saturate the eluent under the chromatographic conditions employed. The precolumn and main column should always be maintained at the same temperature.

Because of the requisite immiscibility, the stationary and mobile phases differ greatly in their solvent properties toward the sample. Consequently, if the sample components are very soluble in the stationary phase, the retention times will be very long and the detection sensitivity low (broad peaks), whereas with high solubility in the mobile phase the retention times will not differ appreciably from that of an unretained sample.

To avoid these problems, ternary mixtures are frequently used. They consist of a nonpolar solvent (e.g., hexane) and a very polar one (e.g., water), to which a third component (e.g., a lower alcohol) is added as a solubilizing agent. Such mixtures reduce the polarity difference between the stationary and mobile liquids, and their compositions can be extensively adjusted to fit a particular separation problem. Since one often works near the miscibility limits of the phases, one must carefully maintain the temperature of the system (reservoir, precolumn, main column, and detector) constant.

A completely inert support for the stationary phase is unknown. The effect of the support and its contribution to the retention as a result of adsorption of the sample on the support surface has already been discussed for gas chromatography [6,7]. These surface adsorption contributions also cannot be neglected for liquid chromatography [8,9], particularly when supports with large surface areas are used.

The support (adsorbent) is in equilibrium with the eluent that is completely saturated with the liquid phase, and adsorbs the latter if it is more polar than the eluent itself, until an equilibrium condition is attained. This can be advantageous because the liquid phase that is lost through mechanical erosion is replaced again [8] up to the equilibrium condition. In this way columns packed with adsorbents may be converted into partition systems. The amount of liquid phase taken up by the support under equilibrium conditions is temperature dependent.

The detector noise is sometimes larger in partition processes than in separations by adsorption because traces of the liquid phase always bleed off the column. This becomes noticeable when a polar solid is coated with a relatively nonpolar liquid phase which is even less polar than the eluent. Making the support hydrophobic, for example by silanizing, reduces such difficulties.

In preparative partition chromatography the isolated sample is contaminated with the stationary phase after removal of the eluent. To avoid this, the system should be selected so that the stationary phase can be easily removed from the sample.

This enumeration of the difficulties associated with partition chromatography should not give the impression that it is unsuitable for HPLC. Numerous successful separations show quite the contrary. Partition chromatography is always to be recommended for the separation of substances that readily undergo catalytic reaction on the surface of active adsorbents. Such activity is largely suppressed by coating with a liquid phase. Partition chromatography is also frequently better suited than adsorption chromatography for the separation of moderately to strongly polar samples. The problem of mechanical erosion of the column can be partially circumvented by the use of chemically bonded stationary phases.

B. Supports and Liquid Phases

1. Supports

Chapter V described supports for the liquid phases. In principle, all porous, adsorptive materials can serve as supports regardless of the size of the specific surface area, of its nature, or the thickness of the porous layer.

Initially, PLB (porous layer beads) were mainly used because of superior column efficiency. However, because of the thin porous layer, they can be coated only with small amounts of stationary phase (maximally with 1 - 2%). Columns packed with PLB impregnated with liquid phase possess all the advantages of these particles: good column efficiency, short analysis times, etc. Because of the relative low coverage, small k' values are obtained. For the same reason, the capacity of such columns is low. The columns show poor stability, as small losses of stationary phase alter the retention volumes substantially.

If totally porous particles (e.g., silica gel), which possess a large pore volume (~ 1 ml/g), are impregnated with so much liquid phase that the pore volume is largely filled, so-called heavily loaded columns are produced that possess certain advantages in liquid chromatography [10]: Column efficiency is higher than that of lightly loaded columns [8]; column capacity is high; peak capacity is also large because the k' values can vary over a wide range; even samples with large k' values are eluted as symmetrical peaks.

Loss of stationary phase through erosion has little effect on the retention volume. The volume available to the mobile phase in the column is smaller than that in columns packed with lightly coated or uncoated supports, so that the volume flow rate required for a given linear flow rate is considerably smaller (cf. II.B).

Both PLB [11] and porous supports with a large surface area [8,9] can adsorb the sample on their surface. For porous supports based on kieselgur (e.g., Chromosorb®) the effect of adsorption on sample retention is small. Such supports possess a large pore volume with a relatively small specific surface area and can be impregnated with large amounts of stationary phase (30 - 50%).

An optimum support for partition chromatography should have the following properties [8]:

Pore volume

The greater the pore volume, the greater is the amount of stationary phase that can be applied before the particles stick to each other. For a pore volume of 1 - 1.3 ml/g the coverage can be easily raised to 1 g liquid phase/g support (called a 100% coverage). The more stationary phase there is on the support and hence in the column, the less important and noticeable are the losses of stationary phase through erosion (column bleed) and the effect on the sample retention volume.

Pore diameter

The average pore diameter of the support has an effect on the band broadening of a retarded sample. For a pore diameter around 40 Å the C-term of an H *vs.* u curve is substantially larger than for a support with a pore diameter of 100 Å or more. Larger molecules may even be excluded at small pore diameters. (In such cases a shorter retention time is observed than that for an unretained sample, as determined with a small, inert molecule; cf. Chapter IX).

Hence the pore diameter determines column stability and its resistance to erosion. Poor column stability in the case of equilibrated phases is due to erosion alone. Because of capillary action the stationary phase is held better by narrow pores than by wide ones. Therefore, columns with narrow-pore supports are stable at considerably higher flow rates than those with wide-pore materials. If the pore diameter is in excess of 1000 Å, the liquid phase will be washed out in a few minutes, even at linear flow rates of less than 5 cm/sec. Columns containing supports with pore diameters of 500 Å or less are stable up to linear flow rates of 10 - 15 cm/sec [9]. Hence, for partition chromatography the pore diameter of the support should lie between 100 - 500 Å.

Specific surface area

The effect of the specific surface area of the support can be ascertained from a comparison of the partition coefficients determined from chromatographic data and by classical partition between two phases. If the support exerts an effect on the retention, the partition coefficients determined chromatographically are always larger than those determined by classical methods. Occasionally, the chromatographically determined values were five to ten times greater than the actual [8]. It was established that for silica gels of varying surface area impregnated with 3,3'-oxydipropionitrile (ODPN) the effect of the support on the retention becomes negligible only if the specific surface area is less than 20 m^2/g [9]. In these cases the same partition coefficient was obtained chromatographically as on silanized Chromosorb®, which is regarded as an inert support. Of course, the effect of the support on sample retention depends also on the structure of the sample. Thus, under identical conditions on a SiO_2/ODPN system a strong dependence on the support was established in the separation of alcohols, whereas in the separation of nitriles only the partition mechanism contributed to the retention [9].

When heavily loaded supports with large specific surface areas reveal the effect of adsorption on sample retention, the amount of stationary phase in the column should be verified. In these cases not only

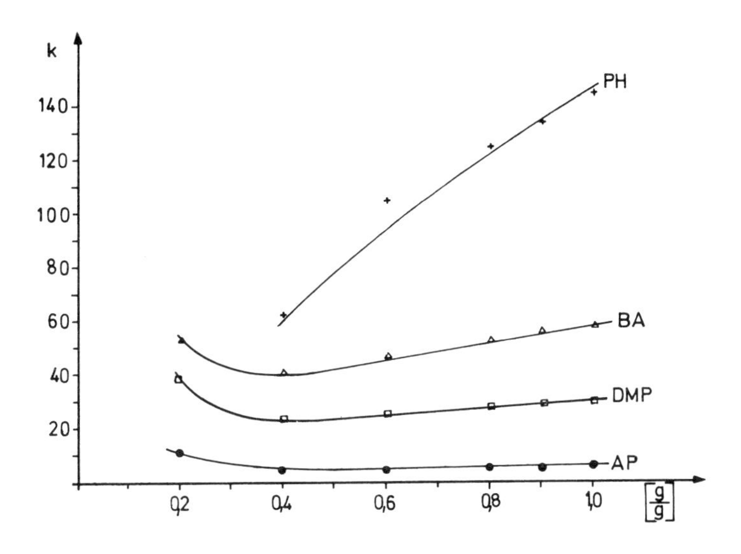

Fig.VII.1. Capacity ratios and the amount of liquid loading: Silica Si 100; liquid phase: 3,3'-oxydipropionitrile; eluent: n-hexane; u = 4 cm/sec. AP = acetophenone; DMP = dimethyl phthalate; BA = benzyl alcohol; PH = phenol

do the absolute retention values change with variation of the surface coverage, but the relative retentions as well. This encumbers the qualitative identification of the chromatographically separated samples. Fig.VII.1 shows the dependence of the k' values of various samples on the surface coverage of a silica gel with a liquid phase. The shape of the curves is typical for mixed adsorption and partition mechanisms [11,12]. Initially, the k' values decrease sharply with increasing coverage of the support with liquid phase, then after passing through a minimum rise again. The decrease in the k' values can be traced back to the elimination of the strongly active adsorption sites. The rise corresponds to an increase in the amount of stationary phase. However, even in this region solid surface adsorption is involved, as can be readily demonstrated by the GC method for the detection of mixed mechanisms [6,7].

In Fig. VII.2 the "partition coefficients" calculated from Fig. VII.1 are plotted against the reciprocal of the liquid phase volume in the column ($1/V_L$). For a "pure" partition mechanism the partition coefficient should be independent of $1/V_L$. For the silica gel used, which has a large specific surface area (500 m^2/g), the contribution from solid adsorption cannot be neglected even for complete coverage with

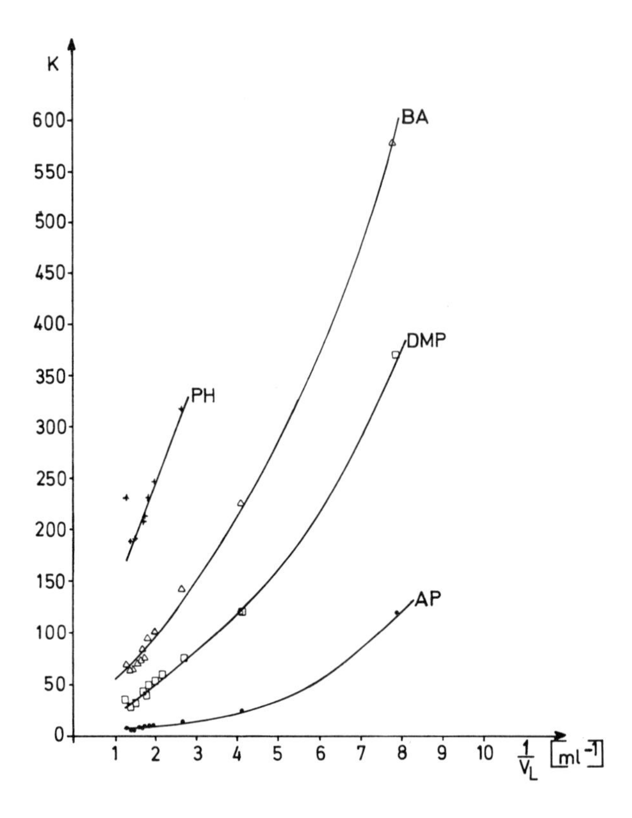

Fig.VII.2. Dependence of the chromatographic partition coefficient on the liquid phase volume V_L. Same conditions as in Fig.VII.1

liquid phase ($1/V_L < 2$). Only for the least retarded substance (AP) is the "chromatographic" partition coefficient, obtained using a heavily loaded column, independent of the amount of stationary phase.

The lifetime of a partition-chromatographic column depends on the resistance of the packing to mechanical erosion. It is assumed that eluents saturated with the liquid phase are used exclusively. As usual, this erosion is small at low eluent flow rates (< 1 cm/sec). However, if one works at higher flow rates, a test mixture with known k' values should be injected periodically to verify the constancy of the k' values and hence of the system.

If supports with large surface areas are employed (in the case of PLB, for example, the silica gel comprising the thin layer has a large surface area [11]) equilibrium is established between the liquid phase adsorbed on the support surface and that dissolved in the eluent, as described in Chapter VI for the adsorbent/water system. This effect can be utilized for the impregnation (see there) of the support with stationary phase. Partition-LLC columns are useful for routine studies only when the impregnation of the support corresponds at least to the equilibrium coverage. Only then are the coverage and hence the retention times and relative retentions independent of the length of service of the column. Obviously, one can operate above the "equilibrium coverage" if high flow rates are avoided, whereas below it continuously varying retention times are encountered.

2. Liquid Phases

Some of the common liquid phases of GC are also suitable for LC and were thus used in the early stages of HPLC (e.g., 3,3'-oxydipropionitrile = ODPN; 1,2,3-tris(2-cyanoethoxy)-propane; trimethylene glycol; polyethylene glycols of different degrees of polymerization, Carbowax®). For such phases, however, only the nonpolar aliphatic hydrocarbons, perhaps with small amounts (max. 10%) of chloroform or tetrahydrofuran and other ethers are suitable mobile phases. All other common eluents are good solvents for these stationary phases. Since the solubility of many organic samples is extremely small in the useable eluents (or mixtures), the applicability of such systems is limited. Furthermore, the differences in selectivity of these stationary phases are small. Fig.VII.3 demonstrates this via the separation of five aromatic alcohols. Polyethylene glycol 400 is best suited for this separation as the stationary phase, but because of its solubility in many eluents the lifetime of such columns is short, even with the use of saturated eluents [13]. Highly viscous liquid phases give poorer column efficiencies than less viscous ones because of low diffusion coefficients [14]. Liquid phases that absorb in the UV cannot be used with UV detectors since even the small amount soluble in the eluent is sufficient to make it opaque to UV radiation.

The choice of an LLC system is simplified because known partition systems from multiple extraction [1] can be applied to HPLC procedures. Here the polar component is loaded on the support and the nonpolar one is used as eluent. In addition, paper chromatographic systems can be applied to HPLC. Thus steroids can be readily separated on paper impregnated with formamide. If the formamide is dispersed on silica gel, the steroids can be separated by HPLC [8] because formamide is practically insoluble even in the moderately polar solvents benzene, methylene chloride, and chloroform.

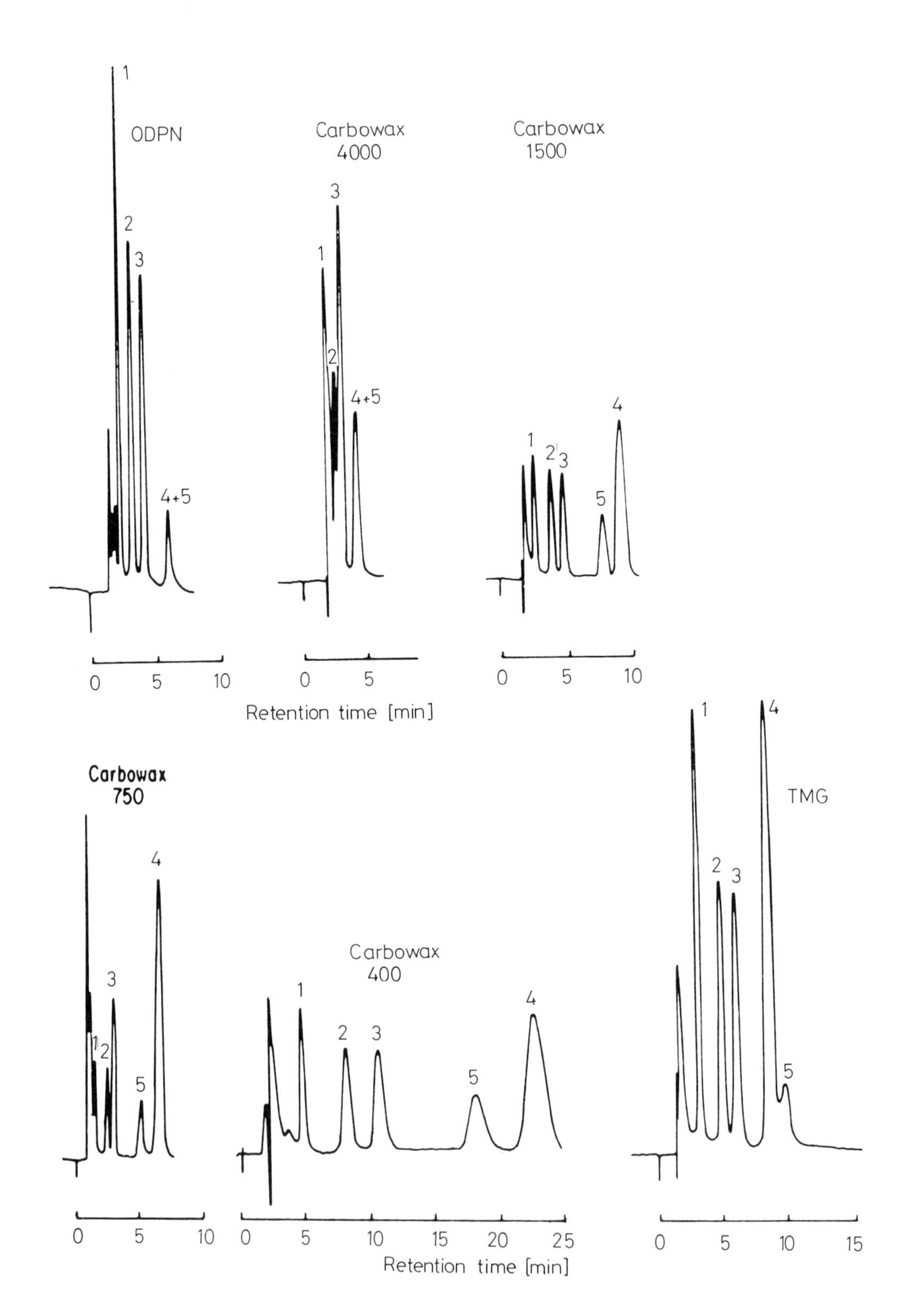

ODPN
1
2
3
4+5
0
5
10
Carbowax
4000
3
1
2
4+5
0
5
Carbowax
1500
4
1
2'
3
5
0
5
10
Retention time [min]
Carbowax
750
4
3
1
2
5
0
5
10
Carbowax
400
1
2
3
5
4
0
5
10
15
20
25
Retention time [min]
TMG
1
4
2
3
5
0
5
10
15

Ternary mixtures [15] should have a far more extensive applicability than the binary systems described thus far. As pointed out already, for simple binary mixtures the two liquids must be very different in order to achieve the required immiscibility. A third component added to such a binary mixture that is completely miscible with both, acts as a solubilizing agent and decreases the polarity difference between the two phases.

Of course, the amount of the third component added must be controlled so as to maintain a two-phase system. If the composition approaches the point of complete miscibility, the selectivity of the system is lost. Furthermore, column stability decreases as the two phases become similar because the mobile phase dissolves the stationary phase. It is also important to select a phase composition that is not unduly affected by small temperature changes. Ternary mixtures of a hydrocarbon (hexane, heptane, isooctane), an alcohol (ethanol, isopropanol) and water have been used. The hydrocarbon may be replaced by more polar solvents such as dichloromethane. Usually the silica is coated with the water-rich phase whereas the more organic one serves as eluent. Prepacked columns must be coated *in situ*. The polar constituents are preferentially adsorbed from the ternary mixture by the stationary phase, especially if its surface area is large, as is detailed in the next section. Plots like those shown in Fig.VII.1, which are typical for partition systems, are obtained.

If ternary mixtures are used with nonpolar stationary phases, the most nonpolar constituent is absorbed by the stationary phase as expected and gives rise to the same phenomena [16] as discussed for polar supports and polar stationary liquids. The selectivity of the system can be altered and adjusted by using different ternary systems with nonpolar stationary phases [17].

Acids and bases may be added to the stationary liquid and (or) to the eluent with polar and nonpolar stationary phases. This may improve the selectivity of the separation of dissociating organic substances. In special cases ion pairs are formed, and these are separated

Fig.VII.3. Selectivity of Liquid Phases [22]. PLB: Zipax, d_p ~ 30 μm, coated with 1% of the liquid phases. ODPN = trimethylene glycol, and Carbowaxes of various mol. wt. Column: 100 cm, 2.1 mm i.d.; eluent: n-hexane; F = 1 ml/min. Samples: 1 = α,α'-dimethylbenzyl alcohol; 2 = α-methyl benzyl alcohol; 3 = phenylethyl alcohol; 4 = cinnamyl alcohol; 5 = benzyl alcohol

with a different selectivity than the individual ions. Ion pair chromatography will be discussed in Section E of this chapter.

3. Coating the Support

The classical methods of coating supports with the liquid phase assume that the stationary phases will be dry-packed. In general, > 25 μm particles are used and are coated with a solution of the liquid phases, the solvent being slowly removed by evaporation. If a suspension technique is employed, the support must be coated in the column after packing, as most liquid phases would be leached out by the suspending medium or during column conditioning.

It is simplest to apply the stationary phase in small amounts while the eluent is flowing through the column. Since small drops of the stationary phase pass through the column and separate out in the detector, etc., a long time is required to attain stable conditions. Whether or not a homogeneous distribution of the liquid phase is achieved is questionable.

A better method, which has been used in gas chromatography, consists of forcing a solution of the desired liquid phase through the column and then displacing the solvent with either a gas (e.g., nitrogen) [18] or an eluent [19]. In either case, the amount of liquid phase taken up by the support cannot be predicted.

If an active solid with a large specific surface area is used as support in conjunction with a polar liquid phase, any coating of the support would be superfluous. In these cases an equilibrium is established between the polar liquid phase dissolved in the eluent and that adsorbed on the active solid. The greater the polarity difference between the stationary and mobile phases, and the greater the specific surface area of the support the more liquid phase is taken up by the solid support. The time required to establish equilibrium depends primarily on the liquid phase concentration in the eluent and its flow rate. It is expedient to add an excess of stationary phase to the eluent reservoir and to recycle the eluent during the coating. A precolumn packed with Chromosorb® impregnated with the desired liquid phase can also be used to saturate the eluent and thus serve as the liquid phase reservoir. However, the precolumn must contain a sufficient excess of liquid phase. The coating time can be shortened substantially by raising the reservoir or precolumn temperature by 1 - 2°C above that of the column.

Such *"in situ"* [20] or *"naturally"* [21] coated columns, having an equilibrium distribution of liquid phase, possess extended lifetimes because any liquid phase dissolved away is continuously replenished as long as the temperature and liquid phase concentration remain constant. Variations in the amount of liquid phase dissolved in the eluent may alter the equilibrium distribution on the support. Since in this case not only do the retention times decrease or increase but the relative retentions change also, the system can be optimized for a particular separation process (compare Fig.VII.9 and 10).

The same column can be coated successively with different liquid phases, if the stationary phase is first removed with a suitable solvent. Before being recoated, the column must be regenerated and activated, i.e., the residual polar wash liquid and perhaps water must be removed with a dry eluent of low polarity. Experience indicates that about 20 column volumes of wash liquid (e.g., methanol) are required to clean a column completely. (Even strongly retained substances are removed by this procedure.) For column regeneration (i.e., activation) 40 - 100 column volumes of a nonpolar eluent (for example, dichloromethane or heptane) are needed.

4. Determination of the Coverage

In the usual coating from solution the amount of liquid phase can be predetermined by weighing. It is assumed that the liquid phase is not volatilized along with the solvent vapor. The determination of the amount of stationary phase in an *in situ* coated column is more difficult. The exact amount of coverage can be determined very accurately by classical C, H-analysis of the stationary phase (support and liquid phase).

If the stationary phase is still in the column, the degree of coverage can be estimated from the total porosity (cf. Chapter II.B). The porosity ε_T of a column packed with uncoated silica gel is always between 0.82 and 0.85. When all pores are filled with stationary phase, the column behaves toward an unretained substance as if it were filled with nonporous glass beads. Column porosity is then 0.42 - 0.45. If the pore volume and packing density of the support in the column are known, the coverage with stationary phase can be calculated.

Table VII.1 compares experimental porosity values with the coverages calculated from them. The results were obtained on a silica gel column (Lichrosorb SI 100) having a large pore volume (1 ml/g).

Table VII.1. Porosity and extent of coating of a silica gel column

Porosity	Coating [g/g] (calculated)	Linear velocity [cm/sec]
0.88	-	0.76
0.84	0.08	0.86
0.80	0.15	0.9
0.76	0.22	0.92
0.70	0.33	0.95
0.44	0.80	1.35

Column: 30 cm, 4 mm i.d.; 2.08 g SiO_2, ®Merckogel Si 100; 10 µm, CH_2Cl_2; F = 5 ml/min; Δp = 100 atm

Water was used as stationary phase. The linear velocity increases to the same extent that the porosity decreases if the volume flow rate is kept constant. A constant pressure drop and volume flow rate are attained only if the support does not swell or shrink during the coating process.

C. Properties of the Column

1. Stability

The *stability* and hence the lifetime of the column depends on the following parameters, which have already been discussed in detail:

a) Saturation of the eluent with stationary phase, which can be conveniently achieved with a more heavily loaded precolumn than the analytical column. Other supports (e.g., kieselgur) and particle sizes can also be used. Changes in the solubility of the liquid phase at higher pressures can be compensated in this manner. (However, such effects are not to be expected at the usual pressures of < 400 atm.)

b) Temperature constancy between precolumn, analytical column, and reservoir, because the composition of the mixtures frequently exhibits a temperature dependence and, occasionally, the amount of liquid phase on the stationary phase changes as a result of temperature variation. With small particles a temperature gradient may appear within the column due to heating of the eluent by the heat of friction [22,23]. At the end of the column there is less liquid phase at higher temperature than at the beginning.

c) Mechanical erosion may remove some liquid phase from the column.

d) Samples should always be dissolved in the column effluent, never in pure unsaturated solvent to avoid stripping the stationary phase.

All of the above effects diminish the lifetime of a column. A decrease in the amount of liquid phase on an inert support always manifests itself by a decrease in the retention times. With active supports the retention times may decrease as well as increase (Fig.VII.1). For "heavily loaded" columns (coverage > 0.3 g liquid phase/g support), small changes in the amount of liquid phase have a less perceptible effect on retention times.

2. Sample Capacity

The sample capacity of LLC systems is an order of magnitude greater than that of pure adsorption systems. Between 10^{-3} and 10^{-2} g of sample/g liquid phase can be applied without appreciably increasing band broadening. However, band broadening increases considerably more rapidly than the k' values decrease on overloaded columns. Fig.VII.4 presents curves obtained with variously loaded supports. If the increase in band broadening is used as a measure of sample capacity, it is found that the sample capacity is independent of the k' value. The overloading of a column (increased band broadening) does not always result in tailing. If changes in the k' values are regarded as the capacity limit, somewhat higher values are obtained. Despite this high capacity, HPLC systems are frequently overloaded, especially when the total amount of liquid phase in the column is small, e.g., with porous layer beads.

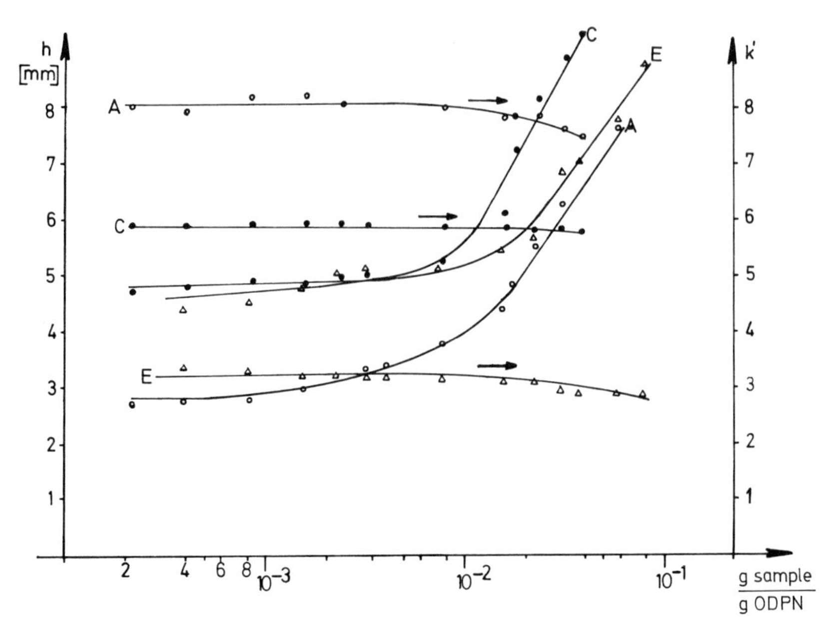

Fig.VII.4. Load capacity of partition systems ("heavily loaded columns") [9]. Stationary phases: A = 0.91 g ODPN/g Porasil A; C = 0.79 g ODPN/g Porasil C; E = 0.34 g ODPN/g Porasil E. All maximally coated. Column: 50 cm, 2 mm i.d.; eluent: n-heptane; u = 1.97 cm/sec; T = 34°C; sample: benzonitrile

3. Preparative Applications

LLC is well suited for preparative separations because of its high sample capacity. Up to 10 mg of each component can be applied onto a 2 mm i.d. column, 50 cm long, that is filled with "heavily loaded" silica gel. This is sufficient for many analytical procedures and detection reactions. If the separation is good, the column can be loaded with even greater amounts of sample.

It should be pointed out again that the isolated sample components remain contaminated with the liquid phase after removal of the eluent. Thus, for preparative work in particular, the liquid phase should be carefully chosen so that it can be easily removed.

4. Column Efficiency

The efficiency of LLC columns depends, of course, on the particle size of the support. Furthermore, the viscosity of the liquid phase affects

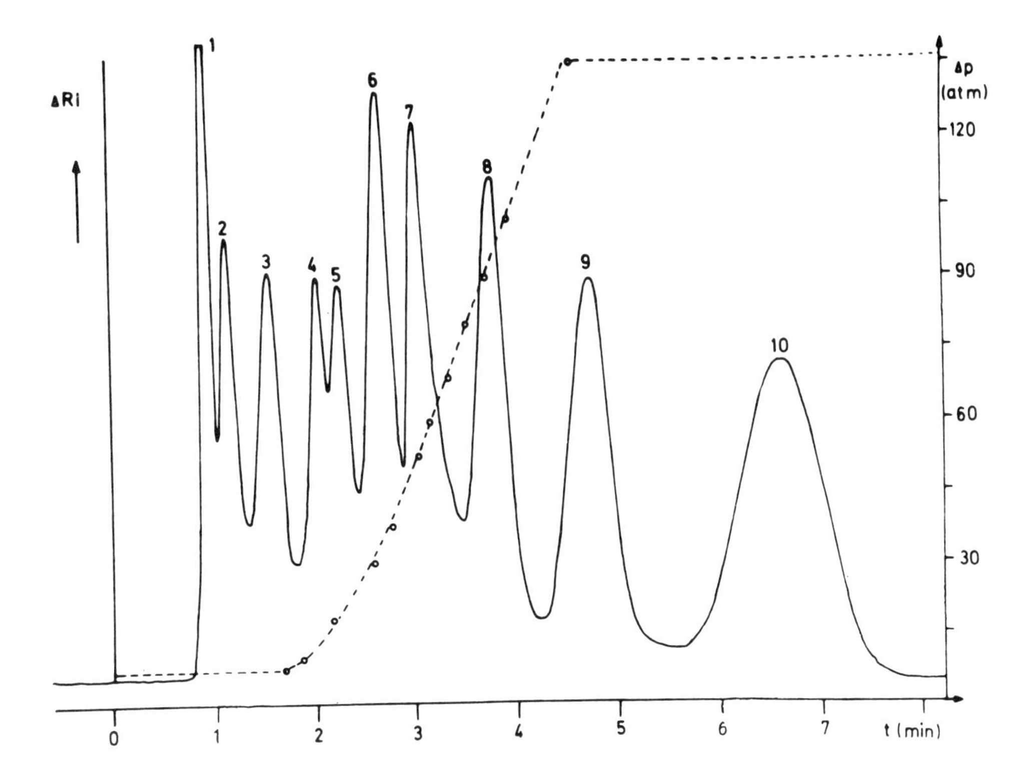

Fig.VII.5. Partition chromatography. Pressure program [9]. Stationary phase: 0.72 g ODPN/g Porasil C esterified with OPN. $d_p \sim 40$ µm. Column: 50 cm, 2 mm i.d.; eluent: n-heptane. Pressure program from p = 7.5 to 135 atm. T = 34°C. 1 = nonane (k' = 0); 2 = thionaphthene (0.3); 3 = N,N'-dimethyl aniline (0.8); 4 = α-naphthaquinoline (1.6); 5 = quinaldine (2.4); 6 = p-benzodiazine (3.95); 7 = isoquinoline (6.5); 8 = phenylpropanol (16.8); 9 = benzyl alcohol (32); 10 = anisyl alcohol (63)

band broadening [11]. The efficiency of "heavily loaded" columns follows the theoretically predicted dependence of the H values (C-term) on the k' value [24]. Maximum H values are obtained at k' values around 1. For k' values over 30, the H values are comparable to those of unretained peaks [8,9]. Surprisingly, this dependence is not always found for small particles (d_p < 10 µm), where the H values may be nearly independent of the k' value of the sample [20].

5. Programming Techniques

Of the programming techniques used in adsorption (LSC) chromatography for reducing the analysis time, only pressure or flow programming can be employed in LLC. A pressure-programmed analysis is shown in Fig.VII.5.

If the separation had been carried out at the low initial flow rate of 0.9 cm/sec, it would have taken more than 60 min.

Temperature programming and gradient elution are not applicable because the composition of the support-liquid phase system changes in both cases or the liquid phase may be completely stripped off.

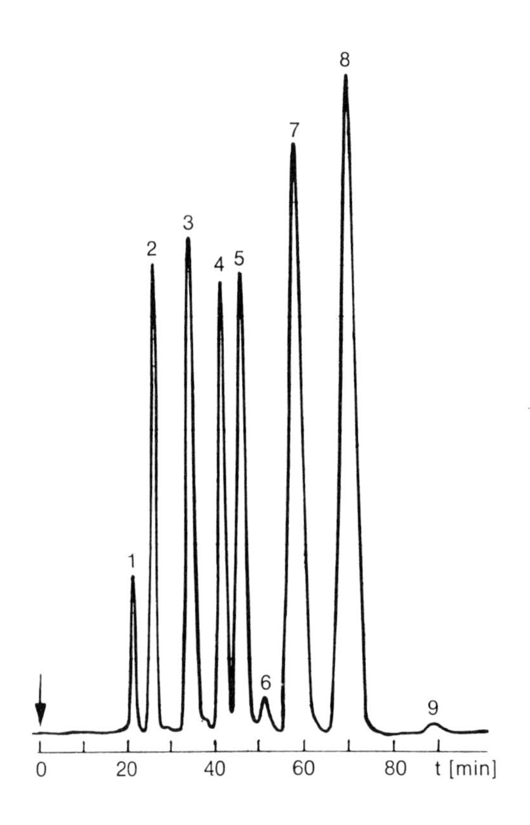

Fig.VII.6. Separation of condensed aromatics (heavily loaded column) (Merck application 72). Stationary phase: Merckosorb Si 60, coated with 50% Fractonitrile III. Column: 250 cm, 2 mm i.d.; eluent: n-heptane; F = 15 ml/h. 1 = benzene; 2 = naphthalene; 3 = anthracene; 4 = pyrene; 5 = fluoranthene; 6 = tetracene; 7 = chrysene; 8 = benzpyrene; 9 = coronene

D. Applications

Liquid-liquid chromatography is best suited for the separation of moderately polar to polar substances, that are too polar for chromatography on silica. The area of application of LLC overlaps that of chromatography with nonpolar stationary phases. This is one reason why the use of LLC has diminished. As with RP systems, nonpolar compounds (such as the condensed aromatic hydrocarbons) can also be separated, as shown in Fig. VII.6. The analysis time is longer than for the separation presented in Fig.VI.23, but this can be attributed to the large particle size. Fast analysis time for the same particle size can be attained with PLB coated with liquid phase. A very rapid analysis of *plasticizers* is shown in Fig.VII.7. This separation could also

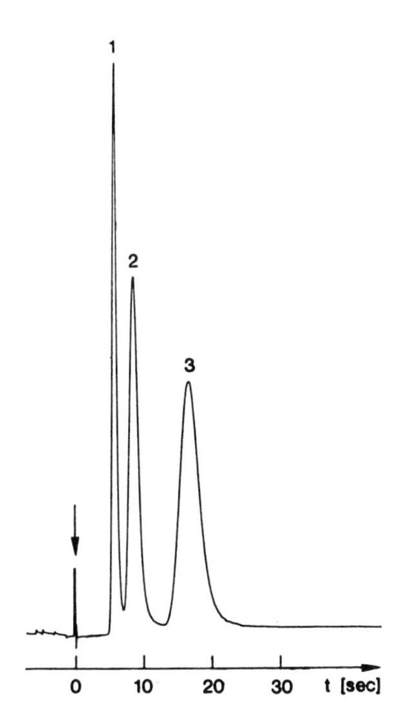

Fig.VII.7. Separation of plasticizers (PLB) (Merck application 73/1). Stationary phase: Perisorb, coated with 1.3% ODPN. Column: 50 cm, 3 mm i.d.; eluent: n-heptane; Δp = 75 atm; F = 6.7 ml/min. 1 = dibutyl phthalate; 2 = diethyl phthalate; 3 = dimethyl phthalate

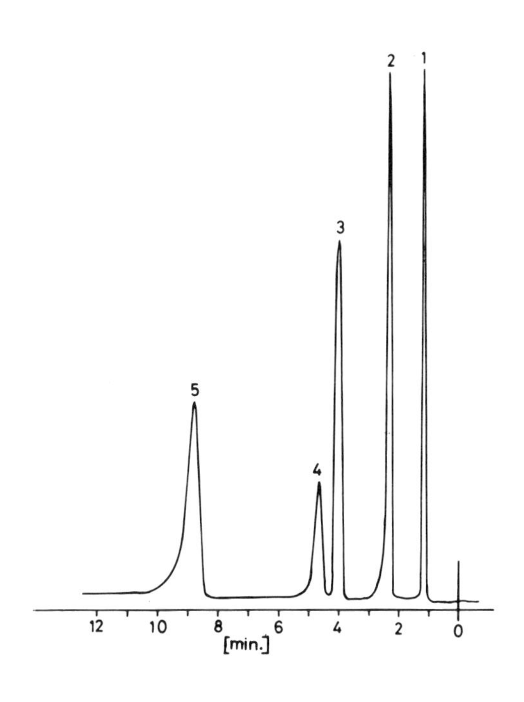

Fig.VII.8. Separation of steroids (heavily loaded column). Stationary phase: Silica Si 100. d_p ~ 10 µm; coated *in situ* with ca. 50% formamide. Column: 30 cm, 4.2 mm i.d.; eluent: methylene chloride; Δp = 105 atm; u = 0.45 cm/sec. 1 = unretained; 2 = corticosterone (k' = 1.3); 3 = cortisone (3.0); 4 = aldosterone (3.6); 5 = hydrocortisone (7.3)

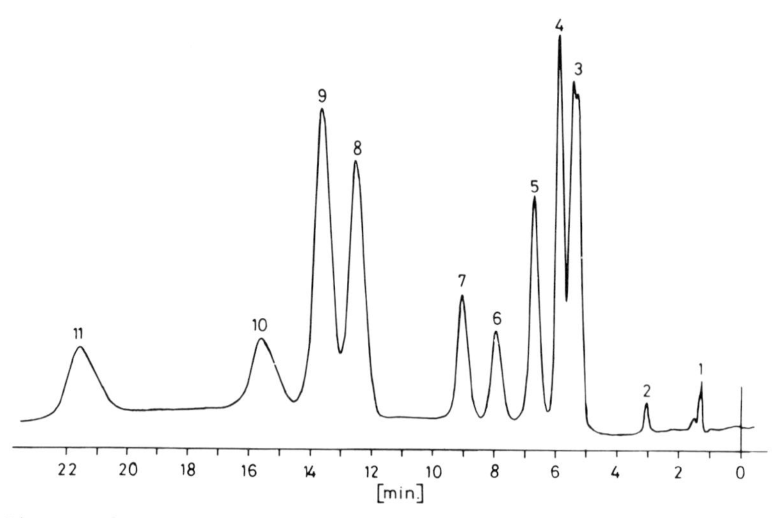

Fig. VII.9. Separation of dansyl amino acids I. Stationary phase: Lichrospher Si 100, d_p ~ 10 µm, coated with ca. 0.4 g/g of the polar component from the eluent. Eluent: methylene chloride (water-saturated) + 1% acetic acid + 1% 2-chloroethanol. Column: 50 cm, 4.2 mm i.d.; Δp = 255 atm; u = 0.6 cm/sec; 1 = unretained; 2 = unknown; 3 = dansyl-isoleucine (k' = 2.9); 4 = dansyl-valine (3.25); 5 = dansyl-leucine (3.9); 6 = dansyl-tyresine (4.7); 7 = dansyl-alanine (6.5);8 = dansyl-tryptophane (8.0); 9 = dansyl-glycine (8.8); 10 = dansyl-histidine (10.1); 11 = dansyl-lysine (14.4)

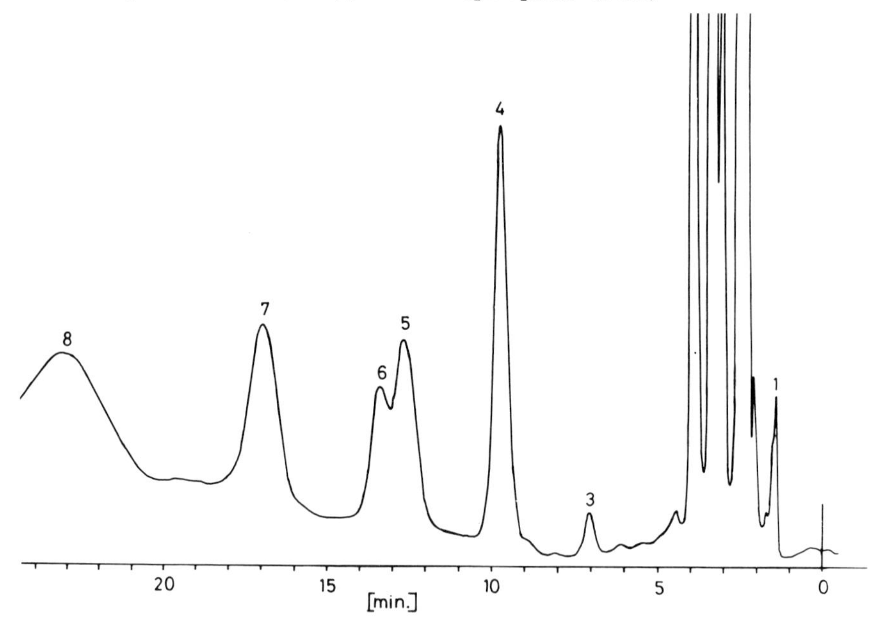

Fig.VII.10. Separation of dansyl amino acids II. [20] Stationary phase and conditions as in Fig.VII.9, except that the eluent contains 10% 2-chloroethanol. 1 = unretained, mixture from Fig.VII.9; 3 = unknown; 4 = dansyl-threonine (k' = 5.9); 5 = dansyl-serine (8.0); 6 = dansyl-glutamic acid (8.5); 7 = dansyl aspargic acid (11.0); 8 = dansyl-cystine (15.5)

have been readily performed by LSC. However, polar classes of compounds, such as *steroids*, are better separated by LLC. Fig.VII.8 shows the separation of some *corticosteroids* with the system silica gel-formamide-methylene chloride [8,20]. Other systems [25-32] have been employed successfully for the separation and determination of steroids. In addition *pesticides* [33], *phenols* [8,14,20], *phenol carboxylic acids* [34], nonionic *surfactants* [35], *brain gangliosides* [36], and *metal chelates* [37] have been separated by HPLC.

Amino acid derivatives, e.g., the *phenylthiohydantoins* [21] formed in the Edman degradation of peptides, and the *dansyl amino acids* [20] have been successfully separated. Figs. VII.9 and VII.10 illustrate the separation of the dansyl derivatives of the most important amino acids. The liquid phase was adsorbed by the support (silica gel) from the mobile phase. A mixture of water-saturated methylene chloride with 1% each of glacial acetic acid and 2-chloroethanol was best suited as eluent for the separation of the derivatives of the monofunctional amino acids (Fig.VII.9). Derivatives of the polar amino acids are excessively retained in this system. After raising the 2-chloroethanol content to 10%, even the dansyl derivatives of the polar amino acids can be separated on the same column after equilibration (Fig.VII.10). The nonpolar amino acid derivatives, however, elute in the vicinity of the unretained peak.

E. Ion-pair Chromatography

The separation of samples that dissociate in aqueous solution can be optimized by adjusting the pH of the aqueous stationary or mobile phase to suppress the dissociation, thereby eluting them as sharper zones. In paper and thin-layer chromatography glacial acetic acid or dilute HCl is added to the solvent in the separation of acids to achieve sharper peaks. For bases, an analogous procedure is employed by adding ammonia or weak organic bases to the developer.

Moreover, suitable counter-ions may be added to the stationary or mobile phase to promote "salt formation" between them and the acidic or basic samples. The "ion pairs" formed alter the retention behavior of ionic compounds substantially, whereas that of nonionic substances is not affected. Hence, an additional parameter is available for the optimization of a separation.

"Ion-pair chromatography" is based on the systematic work of Schill [38,39] and has found extensive application in HPLC due to the lack of suitable ion exchangers for the separation of ionic compounds. Various designations are used for this method, depending on the phase systems utilized, but the fundamental principle is the same in all cases.

In ion-pair chromatography silica gel serves as the support for the aqueous stationary phase, which contains the counter (e.g., perchlorate) and sometimes the necessary buffer. The eluent is immiscible with water [40-43]. On the other hand, in paired-ion chromatography, PIC® [44] a nonpolar phase (RP) is used as stationary phase. For the separations of acids, an organic base (e.g., tetrabutylammonium phosphate) is added to the eluent (e.g., aqueous methanol); for bases an organic acid (e.g., 1-heptane sulfonic acid) is used [45-49]. In the so-called "soap chromatography" [50] organic counter-ions with long carbon chains ($> C_{10}$) are employed.

In ion-pair separations the extent of dissociation of the sample and counter-ion, as well as the corresponding ion-pair formation, may be varied by adjusting the pH of the stationary or mobile phase. The system can thus be made very selective with respect to ion-pair formation, thereby optimizing the separation of the components of interest. The possibilities will be illustrated, using a carboxylic acid as an example:

The dissociation of the carboxylic is based on the following equilibrium:

$$RCOOH \rightleftharpoons RCOO^{\ominus} + H^{\oplus} .$$

By adding an acid (or an acidic buffer) the equilibrium can be displaced to the left, i.e., the dissociation can be repressed. This leads to sharper elution zones that may show slight tailing at most. The dissociation of strongly acidic samples, e.g., those that are completely dissociated at pH 2, cannot be repressed by the addition of acids, in part also because of instrumental limitations (corrosion, etc.). Such samples are scarcely retarded, partly excluded, or eluted as strongly asymmetric peaks. However, the addition of a suitable counter-ion, such as a quaternary ammonium salt, results in the formation of ion pairs for which the partition coefficient is, of course, different from that of the free acid. In addition to acid dissociation, the following equilibria are also involved in ion-pair formation:

The dissociation equilibrium of the added counter-ion

$$R_4NCl \rightleftharpoons R_4N^{\oplus} + Cl^{\ominus}$$

and the formation of the ion-pair

$$R_4N^{\oplus} + RCOO^{\ominus} \rightleftharpoons \boxed{RCOO^{\ominus}\ NR_4^{\oplus}}\ .$$

Since the extent of ion-pair formation depends on the pH of the stationary or mobile phase, it should be adjusted so that the sample and counter-ion are extensively dissociated. The counter-ions used should remain completely dissociated over a wide pH range so that the pH can be adjusted to the optimal value for the separation. Therefore, in practice strong acids (perchloric, alkylsulfonic) and salts of strong bases (quaternary ammonium salts) are used. The pH is held constant in the range of 2 to 8 by means of buffer solutions. At higher or lower pH values technical problems may arise, such as corrosion of the frits, dissolution of the silica gel, etc. The position of the equilibrium and the rate of ion-pair formation depends on the nature of the sample, the type and concentration of the counter-ion, and the pH value. These provide additional variables for the optimization of a separation, besides the usual chromatographic parameters such as the type of stationary phase and the polarity of the eluent. The selectivity can be further enhanced by varying the temperature. Analogous considerations hold for the separation of bases.

In practice, ion-pair chromatography using reversed phases appears to be simpler because the counter-ion can be added to the eluent.

In Fig.VII.11 the separation of different arylsulfonic acids on a nonpolar stationary phase is shown. Tetrabutylammonium phosphate (0.005 M) was added to the eluent (methanol-water 1:1). Without the PIC® reagent the sample components would have been eluted almost unseparated and partially excluded, i.e., eluted before the inert sample D_2O. This exclusion effect can be diminished by the addition of neutral salts, such as NaCl, to the eluent. However, the selectivity for this separation is best with ion-pair formation. It appears to be more difficult to introduce the counter-ions into the aqueous stationary phase of columns already packed with silica gel, to be used with organic mobile phases. The injection method - see Section B.3 - seems to be the only possible way to form an aqueous stationary liquid phase with dissolved counter-ions such as perchlorate, sulfate, etc. The eluent in these cases consists of dichloromethane, usually containing an alcohol (such as pentanol) saturated with the aqueous stationary liquid phase. The formation of ion pairs and that these are the eluting

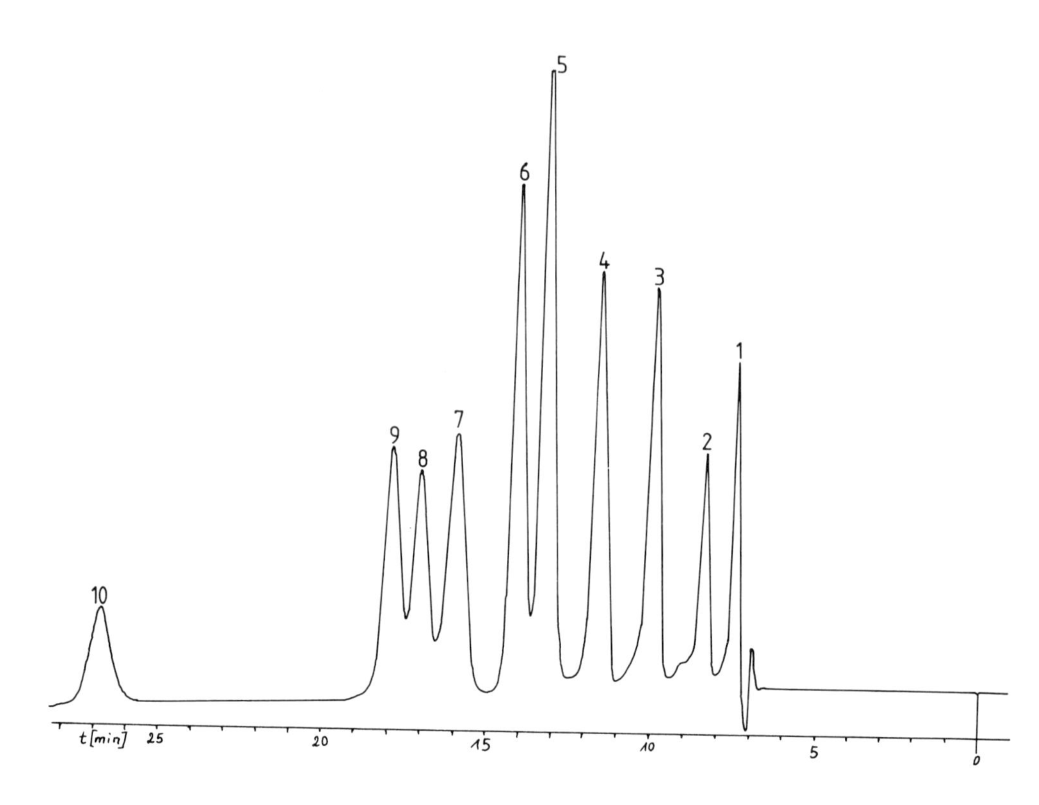

Fig.VII.11. Ion-pair chromatographic separation of sulfonic acids. Stationary phase: RP C_{18} on Silica Si 100; column: 30 cm, 4.2 mm i.d.; eluent: methanol-water (1:1) 0.0005 M tetrabutylammonium phosphate; u = 0.7 mm/sec; Δp = 35 atm; samples: 1 = sulfanilic acid (k' = 0.1); 2 = 2-amino-6-naphthalene-sulfonic acid (0.21); 3 = 2-hydroxy-3,6-naphthalene-disulfonic acid (0.44); 4 = p-toluene-sulfonic acid (0.67); 5 = 1-amino-8-naphthalene-sulfonic acid (0.91); 6 = nitrotoluene-2-sulfonic acid (1.04); 7 = 1-hydroxymethyl-2,5-dihydroxynaphthalene-6-sulfonic acid (1.29); 8 = 1-naphthalene-sulfonic acid (1.44); 9 = 2-naphthalene-sulfonic acid (1.57); 10 = 2-hydroxy-1-naphthalene-sulfonic acid (2.93)

species was elegantly demonstrated by the ion-pair formation of naphthalene sulfonic acid with aliphatic amines [51]. Naphthalene sulfonic acid was coated onto silica in the stationary liquid phase, the aliphatic amines were the samples. Only the ion pairs formed during the separation with chloroform-pentanol as mobile phase are eluted and detected with a UV detector.

Ion-pair chromatography has been applied, for example, to the separation of biogenic amines and their metabolic products [41,52], pharmaceuticals [42,53], carboxylic acids [48], ascorbic acid [49], and dye intermediates [50].

Inorganic ions (lanthanides, actinides, etc.) have been separated with di-(2-ethylhexyl)-phosphoric acid dissolved in dodecane coated as stationary liquid onto silanized silica [54]. The elution was carried out stepwise with nitric acid, hydrochloric acid, etc.

References Chapter VII

1. Hecker, E.: Verteilungsverfahren im chemischen Laboratorium. Weinheim: Verlag Chemie
2. Martin, A.J.P., Synge, R.L.M.: Biochem. J. *35*, 1358 (1941)
3. Littlewood, A.B.: Gas Chromatography, 2nd ed. New York: Academic Press 1970
4. Leibnitz, E., Struppe, H.G. (Hrsg.): Handbuch der Gas-Chromatographie. Weinheim: Verlag Chemie 1970
5. Locke, D.C., in: Giddings, J.C., Keller, R.A. (Eds.): Advanc. Chromatography. Vol. 8. New York: Dekker 1969
6. Conder, J.R., Locke, D.C., Purnell, J.H.: J. Phys. Chem. *73*, 700 (1969)
7. Cadogan, D.F., Conder, J.R., Locke, D.C., Purnell, J.H.: J. Phys. Chem. *73*, 708 (1969)
8. Engelhardt, H., Weigand, N.: Anal. Chem. *45*, 1149 (1973)
9. Rössler, G., Halász, I.: J. Chromatogr. *92*, 33 (1974)
10. Halász, I., Engelhardt, H., Asshauer, J., Karger, B.: Anal. Chem. *42*, 1460 (1970)
11. Karger, B.L., Engelhardt, H., Conroe, K., Halász, I., in: Stock, N. (Ed.): Gas Chromatograpy 1970. London: Institute of Petroleum 1971
12. Halász, I., Wegner, E.E.: Nature *189*, 570 (1961)
13. Schmit, J.A., in: Kirkland, J.J. (Ed.): Modern Practice of Liquid Chromatography. New York: Wiley 1971
14. Karger, B.L., Conroe, K., Engelhardt, H.: J. Chromatogr. Sci. *8*, 242 (1970)
15. Huber, J.F.K.: J. Chromatogr. Sci. *9*, 72 (1971)
16. Kraak, J.C., Bijster, P.: J. Chromatogr. *143*, 499 (1977)
17. Bakalyar, S.R., McIlwrick, R., Roggendorf, E.: J. Chromatogr. *142*, 353 (1977)
18. Majors, R.E.: Anal. Chem. *45*, 755 (1973)
19. Kirkland, J.J., Dilks jr., C.H.: Anal. Chem. *45*, 1778 (1973)
20. Engelhardt, H., Asshauer, J., Neue, U., Weigand, N.: Anal. Chem. *46*, 336 (1974)
21. Frank, G., Strubert, W.: Chromatographia *6*, 522 (1973)
22. Kirkland, J.J., in: Kirkland, J.J. (Ed.): Modern Practice of Liquid Chromatography. New York: Wiley 1971
23. Halász, I., Asshauer, J., Endele, R.: J. Chromatogr. *112*, 37 (1975)
24. Dal Noggare, S., Juvet, R.S.: Gas-Liquid Chromatography. New York: Interscience 1962.
25. Huber, J.F.K., Meijers, C.A.M., Hulsman, J.A.R.J.: Anal. Chem. *44*, 111 (1972)
26. Hesse, Chr., Hövermann, W.: Chromatographia *6*, 345 (1973)
27. Siggia, S., Dishman, R.A.: Anal. Chem. *42*, 1223 (1970)

28. Karger, B.L., Berry, L.V.: Clin. Chem. *17*, 757 (1971)

29. Meijers, C.A.M., Hulsman, J.A.R.J., Huber, J.F.K.: Z. Anal. Chem. *261*, 347 (1972)

30. Hesse, Chr., Pietrzik, K., Hötzel, D.: Chromatographia *10*, 256 (1977)

31. van den Berg, J.H.M., Mol, Ch.R., Deelder, R.S., Thijssen, J.H.H.: Clin. Chim. Acta *78*, 165 (1977)

32. van den Berg, J.H.M., Milley, J., Vonk, N., Deelder, R.S.: J. Chromatogr. *132*, 421 (1977)

33. Koen, J.G., Huber, J.F.K., Poppe, H., Den Boef, G.: J. Chromatogr. Sci. *8*, 192 (1970)

34. Hövermann, W., Rapp, A., Ziegler, A.: Chromatographia *6*, 317 (1973)

35. Huber, J.F.K., Kolder, F.F.M., Miller, J.M.: Anal. Chem. *44*, 105 (1972)

36. Tjaden, U.R., Krol, J.H., van Hoeven, R.P., Oomen-Meulemans, E.P.M., emmelot, P.: J. Chromatogr. *136*, 233 (1977)

37. Huber, J.F.K., Kraak, J.C., Veening, H.: Anal. Chem. *44*, 1554 (1972)

38. Schill, G.: Acta Pharm. Sue. *2*, 13 (1965)

39. Schill, G., Modin, R., Persson, B.A.: Acta Pharm. Sue. *2*, 119 (1965)

40. Eksborg, S., Schill, G.: Anal. Chem. *45*, 2092 (1973)

41. Karger, B.L., Persson, B.A.: J. Chromatogr. Sci. *12*, 521 (1974)

42. Karger, B.L., Su, S.C., Marchese, S., Persson, B.A.: J. Chromatogr. Sci. *12*, 678 (1974)

43. Kraak, J.C., Huber, J.F.K.: J. Chromatogr. *102*, 333 (1974)

44. Trade name. Waters Associates, Milford, U.S.A.

45. Wittmer, D.P., Nuessle, N.O., Haney, W.G.: Anal. Chem. *47*, 1422 (1975)

46. Wahlund, K.-G.: J. Chromatogr. *115*, 411 (1975)

47. Wahlund, K.-G., Lund, U.: J. Chromatogr. *122*, 269 (1976)

48. Fransson, B., Wahlund, K.-G., Johannsson, J.M., Schill, G.: J. Chromatogr. *125*, 327 (1976)

49. Sood, S.P., Sartori, L.E., Wittmer, D.P., Haney, W.G.: Anal. Chem. *48*, 796 (1976)

50. Knox, J.H., Laird, G.R.: J. Chromatogr. *112*, 17 (1976)

51. Krommen, J., Fransson, B., Schill, G.: J. Chromatogr. *142*, 283 (1977)

52. Persson, B.A., Lagerström, P.-O.: J. Chromatogr. *122*, 305 (1976)

53. Knox, J.H., Jurand, J.: J. Chromatogr. *110*, 103 (1975)

54. Horwitz, E.P., Bloomquist, C.A.A., Delphin, W.H.: J. Chromatogr. Sci. *15*, 41 (1977)

Chapter VIII

Ion-Exchange Chromatography

A. Principle

Ion exchangers consist of an insoluble framework (matrix) containing covalently bonded dissociable functional groups at accessible sites. These are either sulfonic acid groups or, less importantly, carboxyl groups in the case of cation exchangers, and tertiary amino or quaternary ammonium groups for anion exchangers. Substances that are at least partially ionic in strongly polar eluents can be separated by this method. Separation is based on the affinity differences of ions toward their counter-ions in the ion exchange matrix and those dissolved in the eluent.

In the separation of organic ions sorption on the exchange matrix may contribute to the reaction. Thus, the matrix may act as a "reversed-phase" sorbent and affect the elution of these ions.

Familiarity with classical ion exchange chromatography will be assumed [1-5], and therefore only a short description will be presented. Ion exchange can be described simply in terms of the ion exchange equilibria:

Cation exchange:

$$X^{+} \text{ (eluent)} + Y^{+} \text{ (matrix)}^{-} \rightleftharpoons Y^{+} \text{ (eluent)} + X^{+} \text{ (matrix)}^{-}$$

Anion exchange:

$$X^{-} \text{ (eluent)} + Y^{-} \text{ (matrix)}^{+} \rightleftharpoons Y^{-} \text{ (eluent)} + X^{-} \text{ (matrix)}^{+}$$

The separation of cation or anion mixtures is based on differences in the sorption selectivity of the sample ions and those bound to the matrix. Changes in the pH of the eluent may affect the dissociation of the bound or dissolved ions. For example, the separation of mixtures of acids or bases with different pK values can be carried out successfully using a pH gradient. For such separations, however, only strongly acidic or basic ion exchangers should be employed, because the dissociation of weak ones can be so easily suppressed that ion exchange may cease (see the positions of H^+ and OH^- as eluents in Section D.3). The ion exchange equilibrium can also be displaced by changing the ionic strength of the buffer solution. Such change has a far greater effect on the retention volume than a change in pH. Furthermore, the ion exchange equilibrium is affected by changes in the type of buffer used or by addition of suitable complexation reagents.

This can be illustrated by a simple example [6]:

As a rule, multiply-charged ions are bound more strongly than singly-charged ones. The binding strengths of cations (in dilute solution) fall into the order $M^{4+} > M^{3+} > M^{2+} > M^+$. Thus, if Fe^{+3} and Cu^{+2} are bound to an ion exchanger, dilute hydrochloric acid would elute the copper first and then the iron. However, a poor separation of the cations is obtained by displacement with a large excess of protons. On the other hand, with dilute phosphoric acid as eluent, the proton concentration is too low to displace the cations from the exchanger. Iron is eluted, however, because it forms a negatively charged complex with phosphoric acid that is not retained by a cation exchanger. Copper can then be eluted with dilute HCl.

The tendency of various heavy metal ions to form anion-exchangeable halo complexes in concentrated hydrohalic acid solution is frequently used for their separation. The ions are then eluted by continuous or discontinuous reduction of the acid concentration. As the complexes decompose with the formation of the cations, which of course are not retained by an anion exchanger, the metals are eluted from the column.

Ionic complexes of neutral molecules, such as the boric acid complexes of the sugars or the bisulfite addition products of carbonyl compounds, can also be sparated by ion exchange.

Separations can also be carried out by means of ligand exchange reactions on ion exchangers loaded with metal ions [7]. Since only the complexation tendency of metal ion bound to the ion exchangers is used in the separations (for example, of amino acids on exchangers loaded with copper ions), this is as little a matter of "real" ion exchange chromatography as is the separation of polar organic molecules (e.g., sugars) with aqueous alcohol eluents. In the latter case a partition system is formed between an aqueous phase in the exchanger's resin and an aqueous alcohol mobile phase [8].

In addition to the above exchange mechanisms, sorption effects of the nonpolar matrix may contribute to retention, especially in the separation of organic compounds. It is therefore very difficult to predict the selectivity for separations of organic compounds on ion exchangers.

B. Ion Exchange Materials

The commercially available ion exchange materials have already been mentioned in Chapter V. For applications in HPLC, they must, above all, be pressure stable and also possess the properties necessary for classical, low-pressure applications such as insolubility, chemical stability, etc.

Good chromatographic properties (low H values, rapid mass transfer) can only be attained by reducing the diffusion distances. This can be achieved either by decreasing the particle diameter or by using PLB particles. The following types of ion exchangers have been applied in HPLC:

1. Ion Exchangers with an Organic Polymer Matrix

These are resins with a small particle size (5 - 20 µm), which are also used in classical ion exchange chromatography. Ion exchangers with a higher degree of cross-linking (polystyrene resins containing 6 - 8% divinylbenzene) are utilized primarily because of their lower swelling capacity and better pressure stability. In addition, spherical ion exchangers with a particle diameter of 5 - 10 µm are prepared especially for HPLC.

In using this type of ion exchanger in HPLC it is most important to note that above a certain pressure range the linear eluent velocity fails to increase when the pressure drop along the column is raised. If this occurs, the pressure stability of the matrix has been exceeded. Frequently, column efficiency deteriorates as well. A reduction in pressure does not necessarily restore the original conditions (efficiency, permeability, etc.).

The exchangers are packed into a column in a pre-swollen state. Of course, changes in the ionic strength or pH alter the volume of the swollen exchanger, which may affect column permeability as well as efficiency (e.g., by formation of cavities). These ion exchangers have a large capacity (several μeq/g).

2. Polymeric Ion Exchangers on PLB Particles

Distinction is made between two groups:

a) Pellicular ion exchangers

In this case a film of a polymeric organic ion exchanger is applied onto (~ 30 μm) solid glass beads [9] yielding a layer about 1 μm thick. Since the exchanger layer does not need to be pressure-stable, slightly cross-linked polymers can be used. This contributes, as does the thin exchanger layer, to improved chromatographic properties, especially the speed of analysis. A further advantage is that these exchangers can be dry-packed. The bed volume is independent of the ionic strength of the eluent.

b) Superficially porous ion exchangers

This group does not differ appreciably from the pellicular ion exchangers. The polymeric organic ion exchanger is applied onto a glass bead covered with a thin layer of porous silica gel [10]. Because the quantity of ion exchanger is small in this case as well as in the preceding, the exchange capacity is low (~ 10 μeq/g). Only small quantities can be separated on a column. Hence very sensitive detectors are required. Because the polymer exchange layer is soluble in organic solvents, such ion exchangers are limited to purely aqueous systems.

3. Brush Type Ion Exchangers

Not only can polymers be coated onto PLB, but organic residues of the "brush" type can be covalently bonded as well, and ion exchange groups can then be introduced. The organic residues can be bound as alkyl or aryl groups. Such products are commercially available.

Because of the small surface area of the PLB supports the exchange capacity is small (30 - 60 μeq/g). Organic eluents can be used with such exchangers. On the other hand, if one starts with completely porous silica gels having a large surface area, and carries out the appropriate reactions (silanizing followed by introduction of the ion exchange groups), one obtains ion exchangers with a larger capacity (200 - 500 μeq/g) [11,13]. The exchange capacity is proportional to the specific surface area of the silica gel. With sufficiently small particle diameters (5 or 10 μm) excellent column efficiencies and fast analyses are obtained. These exchangers are not compressible and changes in the eluent composition (e.g., variations in ionic strength) do not affect their bed volume. Since the silica gel support begins to dissolve above pH 8.5, they cannot be used in the strongly alkaline region.

4. Liquid Ion Exchangers

Of course, silica gel can also be coated with so-called "liquid ion exchangers" as in partition chromatography. The "liquid ion exchangers" involve a) water insoluble (long-chain) tertiary amines or quaternary ammonium compounds, that are used as anion exchangers, and b) liquid cation exchangers, such as the dialkyl esters of phosphoric acid. The stability is low because of the relatively low cohesive strength between the "liquid ion exchanger" with its nonpolar organic residue and the polar silica gel surface. Many liquid ion exchangers are surfactants which tend to form emulsions in water. Making the silica gel surface hydrophobic (by silanizing) improves the adhesive strength of the coating. The advantages and disadvantages of these systems are similar to those described in Chapter VII for partition systems. Radioactive isotopes have been separated by liquid ion exchange on a silica column coated with a dialkylsulfonic acid in dodecane. Nitric acid was used as eluent [14].

It appears that for ion-pair formation [15-17] the application of such systems is being extended to the separation of ionic compounds: whether the separation is based on partition or ion exchange may be very difficult to determine. Ion-pair chromatography is discussed in detail in Chapter VII.E.

C. Characterization of the Ion Exchangers

Ion exchangers are characterized on the basis of their functional groups. Distinction is made between strong and weak acid exchangers, depending on whether sulfonic acid or carboxyl groups are present. The strongly basic ion exchangers involve tertiary amines or quaternary ammonium compounds, whereas primary amines are found in the weakly basic exchangers.

Because of the substantial effect of the matrix on selectivity and chromatographic properties, it is essential to characterize exactly the support of the functional groups. For ion exchangers with a purely organic matrix the degree of cross-linking of the matrix (e.g., % divinyl benzene added for polystyrene resins), as well as its composition (e.g., polystyrene, polymethyl acrylate, etc.) should always be stated. The degree of cross-linking determines the pressure stability and the accessibility to the functional groups. For the purely organic resins distinction is made between those with micro pores (microreticular) and the recently available macro-porous (macroreticular) resins, for which access to the micro pores is facilitated by the macro pores interlacing a particle. In addition to the distinction between pellicular and superficially porous, for PLB ion exchangers it should also be specified whether the organic residues containing the functional groups are physically adsorbed or chemically bonded. The type of organic residue should be known.

The *exchange capacity* is also important. The greater the exchange capacity the greater is the load capacity. Moreover, at higher exchange capacities the salt concentration in the eluent can exceed 0.01 M because the k' values also increase with rising exchange capacity. At low capacities one is forced to work with small ionic concentrations, so that the reproducibility decreases. If the sample contains salt, complete displacement may occur.

The exchange capacity can be determined by the usual ion exchange methods (cf. [2.5]). For a very rapid determination, the direct acid or base titration of the exchangers in their base or acid forms, respectively, is adequate. The strength of the exchanger can be established from the position of the end point. Only for strongly acidic or basic exchangers does the end point occur at pH 7. The titration of the exchanger should be repeated after several loading and regeneration cycles. The capacity should not change.

Fig.VIII.1 shows the titration curve of a strong acid cation exchanger of the brush type (curve 1) (butylsulfonic acid bonded to silica gel). For comparison, a titration curve (curve 2) of the weakly acid "cation exchanger" silica gel is also presented [12].

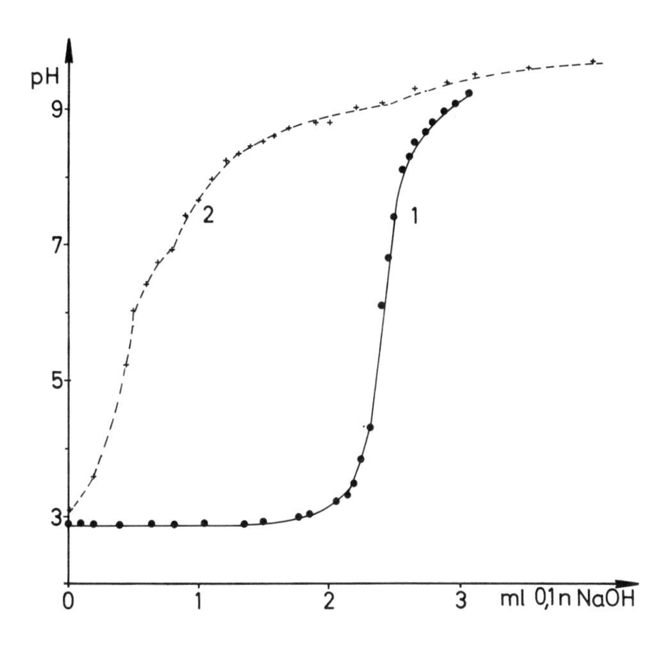

Fig.VIII.1. Titration curves of "bare" silica (2) and chemically bonded cation exchanger (1): n-butylsulfonic acid on silica Si 100.

D. Optimizing a Separation

It is well known that ion exchange equilibria can be described by the law of mass action. Only when additional effects, primarily those of the matrix, play a role in the separation, is it difficult to predict selectivity and elution order.

The properties of the eluent may be responsible for the success or failure of a separation. Since water serves almost exclusively as eluent, separations can be influenced by changes in pH, the type of buffer (type of counter-ion), and the ionic strength. Furthermore, the addition of complexing agents or organic components may alter the selectivity.

1. The Effect of pH on Retention

The retention of weak acids and bases depends on the pH of the eluent, because they exist either as dissociated species separable by ion

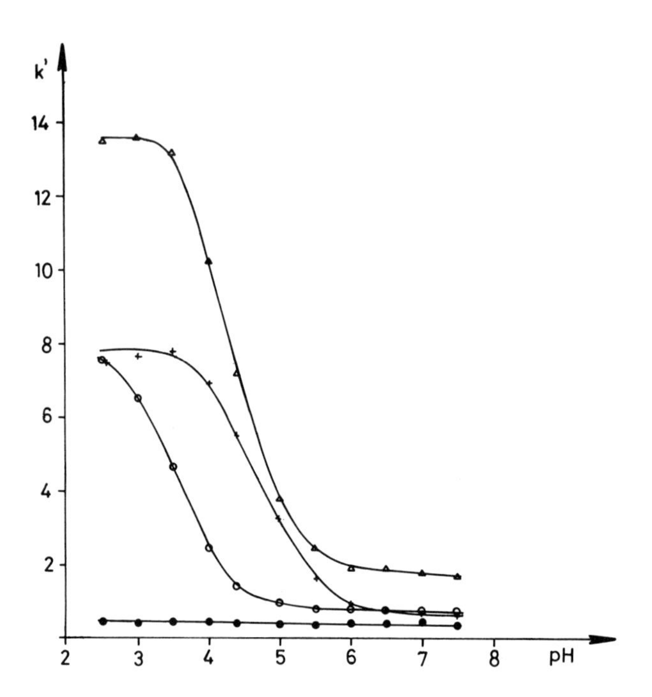

Fig.VIII.2. Effect of pH on k' values (purine and pyrimidine bases). Ion exchanger: n-butylsulfonic acid on silica Si 100; 250 μeq/g; mobile phase: 0.1 M sodium phosphate buffer.
• uracil, o guanine, + cytosine, Δ adenine

exchange or as undissociated molecules not retained by the column. A possible small retention can be attributed to other factors.

The dependence of the retention (k' values) on pH is shown in Fig.VIII.2 for some purine and pyrimidine bases. n-Butylsulfonic acid (bonded to silica gel) served as ion exchanger [12,18]. A 0.1 M phosphate buffer was used as eluent in the pH range of 2.5 - 7.5. Above pH 6.0 the k' values for all bases lie between 1 and 1.8. At this pH the exchanger exists entirely in the Na^+ form, and the free bases cannot compete with the Na^+ ion.

As the pH is decreased the k' values of the base increase. At low pH values the k' values once again become nearly independent of pH because the bases are fully protonated and cannot interact with the ion exchanger. To a good approximation, the inflection point corresponds to the pK value of bases at which they are in a 1 : 1 ratio in equilibrium with their salts. The retention in the pH-independent range can be attributed to a matrix effect. Uracil, which is not ionic in the measured region, is retained only as a result of matrix effects.

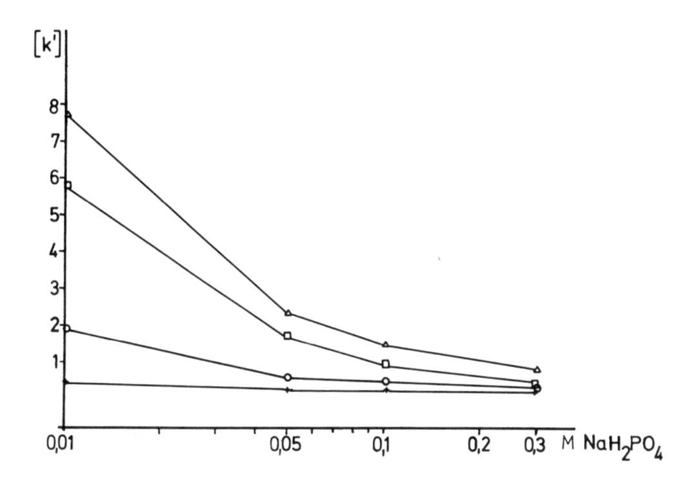

Fig.VIII.3. Effect of ionic strength on k' values (purine and pyramidine bases). Ion exchanger: n-butylsulfonic acid on silica Si 500; 45 μeq/g; eluent: sodium phosphate buffer. Δ adenine, □ cytosine, o guanine, • uracil and thymine

A similar pH dependence of the k' values was found for morphine bases on PLB coated with cation exchanger (®Zipax SCX) in the pH range of 9.1 - 9.8. For this relatively narrow range a plot of log k' = f(pH) yields a straight line [19].

2. Effect of Ionic Strength on Retention

A change in the ionic strength of the eluent affects the retention much more than a change in pH. This fact, known from classical ion exchange chromatography, is illustrated in Fig.VIII.3. For the ion exchanger used, with its relatively low capacity, the largest drop in the k' values occurs when the phosphate concentration is raised from 0.01 M to 0.05 M. For ion exchangers with a higher capacity this region is displaced toward higher ionic concentrations. This effect of the ionic strength on the sample k' values can be ascribed to the displacement of the ion exchange equilibrium. The k' values of the samples are inversely proportional to the ionic strength. If the k' values from Fig. VIII.3 are plotted against the reciprocal concentration, straight lines are obtained that do not pass through zero but intercept at k' values of 0.1 - 0.8. This is also a manifestation of the matrix effects on retention, as the ionic concentration hardly affects reversed-phase sorption on the matrix.

3. Change of the Buffer Solution

A change in the ions of the eluent may affect the selectivity of a separation. Occasionally, a separation is made possible only with the aid of a very special buffer solution. An example is the separation of sugars in the presence of boric acid buffers [20,21], as sugars form ion-exchangeable complexes only with boric acid. With all other buffer solutions the sugars exhibit very small k' values that are attributable to reversed phase effects.

The elution order of anions or cations depends on the strength with which they themselves are retained on a particular exchanger. General rules for the sorption of inorganic ions have already been mentioned. For *anions* the following order has been established on classical, strongly basic anion exchangers:

citrate > oxalate > I^- > HSO_4^- > NO_3^- > Br^- > CI^- > formate > acetate > OH^- > F^-.

This order varies for different commercial products. For weakly basic ion exchangers small shifts among themselves have been established. For example, hydroxyl ions are a strong eluent because the dissociation of the weakly basic anion exchanger diminishes in an alkaline medium.

A similar order has been established for the sorption of cations on strongly acid cation exchangers:

Fe^{3+} > Ba^{2+} > Pb^{2+} > Ca^{2+} > Ni^{2+} > Cd^{2+} > Cu^{2+} > Co^{2+} > Zn^{2+} > Mg^{2+} > UO_2^{2+} > TI^+ > Ag^+ > Cs^+ > Rb^+ > K^+ > NH_4^+ > Na^+ > H^+ > Li^+.

Hydrogen ions are the strongest eluent for weakly acid exchangers which are no longer dissociated below pH 4.5.

In automated amino acid analysis, the change from a sodium citrate to a lithium citrate buffer led to a general increase in the k' values. This enabled the important amino acids, aspargic and glutamic acids, to be separated from each other [22]. It is logical that the selectivity of a separation is altered and the k' values decreased or increased by complexing the sample components with either ions contained in the eluent or with ions bound to the exchanger.

4. Other Effects

Although a temperature rise hardly affects ion-exchange equilibria [5], it can lead to a reduction of the retention times. The reason may lie in a decrease in the non-ion-exchange sorption. At any rate, at higher column temperatures sharper and more symmetric peaks are usually obtained than at lower ones (due to a rise in the diffusion coefficients). As a rule of thumb, a temperature rise from 25 °C to 50 °C doubles the diffusion coefficients. The addition of *organic solvents* (which are completely water-miscible, e.g., the lower alcohols, acetonitrile, tetrahydrofuran) to the eluent changes the selectivity of the ion-exchange separation. A decrease in the hydration of the ions, a change in the dissociation or in the complexation may shift the k' values and the relative retentions. Moreover, column efficiency is raised when eluent viscosity is decreased.

The addition of polar organic components (e.g., alcohols) to the eluent represses ion exchange, and the separation occurs by a partition mechanism. The stationary phase then consists of water which is adsorbed by the bonded ions for their hydration, whereas the eluent is a water-alcohol mixture. Such systems have been employed successfully for the separation of sugars and carbohydrates [cf. 8] as well as of nucleic acids [50].

5. Gradient Elution

Gradient elution is frequently used to elute compounds from ion-exchange columns. Both pH and concentration gradients are utilized. When using ion exchangers with an organic matrix it should be borne in mind that their swollen volume varies with pH, especially on changing the ionic concentration. This becomes evident through changes in the permeability or the deterioration of the column efficiency, if cavities are formed in the packing. Similar phenomena occur during gradient elution when an organic solvent is added to the eluent.

Such problems do not arise with surface-modified silicas or with PLB-based ion exchangers. Although phosphate buffers are supposed to be completely UV-transparent, baseline displacement (at 254 nm) may occur due to the presence of polyphosphate impurities.

E. Applications

In a narrow sense, only the separations carried out on the specially developed, pressure-stable, non-swelling ion exchangers should be cited here. On the other hand, classical ion-exchange separations, such as amino acid analysis, may be shortened substantially [23,24] under HPLC conditions (smaller particle size resins, somewhat higher pressure). The analysis time may be further reduced by replacing the "slow" ninhydrin reaction with ®Fluram (Hoffmann-La Roche), a substance that reacts with primary amino groups to produce an intense fluorescence in the visible region [25,26].

The examples of applications are therefore arranged by the type of ion exchanger used. This is justified because there are already several monographs on ion exchange in HPLC [27-32].

1. Classical Ion Exchangers in HPLC

A separation of amino acids on an ion-exchange resin conforming to HPLC requirements is shown in Fig.VIII.4. The particle size of the unswollen resin was 8 - 9 μm. Sixteen amino acids were separated on a relatively short column (35 cm) and detected with ®Fluram. An eluent flow rate of 0.5 ml/min was attained with a pressure of *ca.* 30 atm.

Such resins, with about 8% cross-linking and a small particle size, can sometimes be subjected to relatively high pressures. C.D. Scott achieved numerous separations on classical organic ion exchangers in determining the organic constituents of body fluids (urine, serum, etc.) [28,33]. In a single urine sample 100 to 120 UV-absorbing substances could be detected; 48 components were identified by means of a carbohydrate-specific reaction [34-36]. Automatic instruments have been developed for clinical laboratories [37]. Sugars have been resolved on organic ion exchangers as borate complexes [38] and by a partition system [39] under HPLC conditions. An example is shown in Fig. VIII.5 [40].

Keto- and hydroxycarboxylic acids have also been separated by HPLC on organic resins [41]. An amino acid analyzer was used to identify aromatic compounds on an ion exchanger [42].

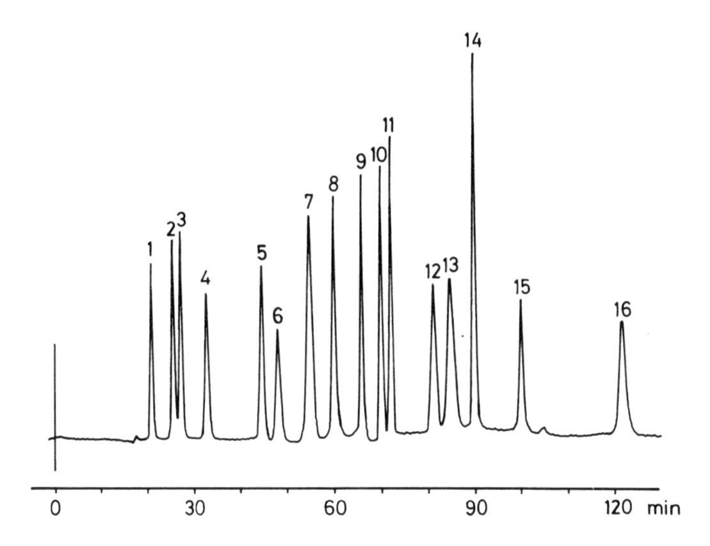

Fig.VIII.4. Amino acid analysis (Durrum data sheet). Ion exchanger: Durrum DC-4A; column: 35 cm, 3.2 mm i.d.; eluent: Durrum Buffer System V. F = 12 ml/h; Δp = 32 atm. Reaction detector: Fluorescamine-Aminco Fluorimeter.
1 = aspargic acid; 2 = threonine; 3 = serine; 4 = glutamic acid; 5 = glycine; 6 = alanine; 7 = cystine; 8 = valine; 9 = methionine; 10 = isoleucine; 11 = leucine; 12 = tyrosine; 13 = phenylalanine; 14 = lysine; 15 = histidine; 16 = arginine

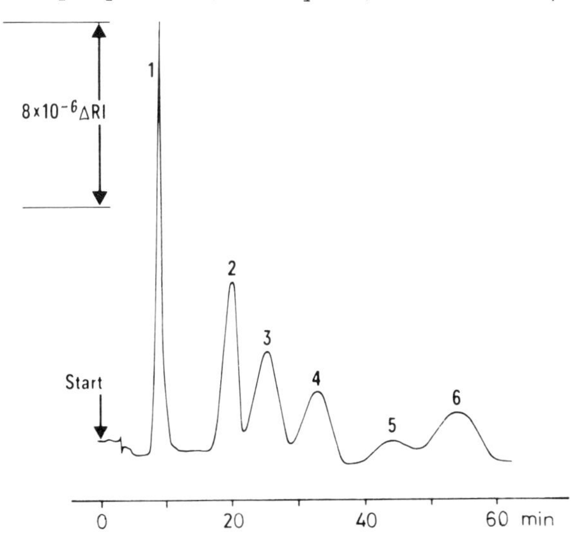

Fig.VIII.5. Partition chromatography on ion exchangers. Separation of sugars (Siemens data sheet 05/05). Ion exchanger: Aminex A-7, Li^+ form; column: 50 cm, 3 mm i.d.; eluent: 85% ethanol, 0.01 M LiCl; F = 0.45 ml/min; Δp = 270 atm; T = 70°C. 1 = rhamnose; 2 = glucose; 3 = saccharose; 4 = trehalose; 5 = melibiose; 6 = raffinose

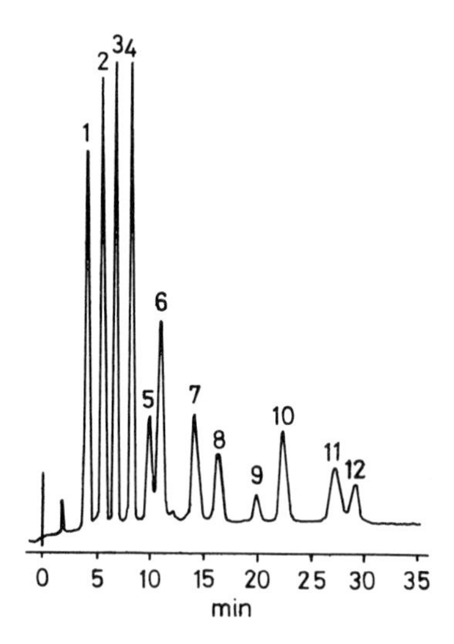

Abb. VIII.6. Separation of nucleotides (Du Pont LC Methods Bulletin 820 M 11). Ion exchanger: Permaphase AAX; column: 100 cm, 2 mm i.d.; F = 1 ml/min; Δp = 70 atm; eluent: exponential gradient from 0.002 M KH_2PO_4 (pH 3.3) to 0.5 M KH_2PO_4. Rate of increase: 3%/min. 1 = CMP, 2 = AMP, 3 = UMO, 4 = GMP, 5 = CDP, 6 = UDP, 7 = ADP, 8 = GDP, 9 = CTP, 10 = UTP, 11 = ATP, 12 = GTP

2. Porous Layer Beads (PLB)

Since the initial work of Horvath and Lipsky [9], numerous separations of nucleic acids and bases have been performed on pellicular ion exchangers [29-32,43,44]. Both cation and anion exchangers have been used; the analysis time is about 30 min. The separation of nucleic acids, the contents of pharmaceutical preparations, etc., has also been successfully carried out on silica gel coated with a polymer film (e.g., ®Zipax SAX) [29,45-47]. Ion exchangers have also proved useful in the detection and separation of drug components (morphine, heroin, methadone) [19].

PLB-based ion exchangers covalently bonded to the silica surface have also been produced. A separation of nucleic acids on such a permanently bonded ion exchanger (e.g., ®Permaphase, DuPont) is presented in Fig.VIII.6. They have also been successfully employed in nucleotide analysis [48].

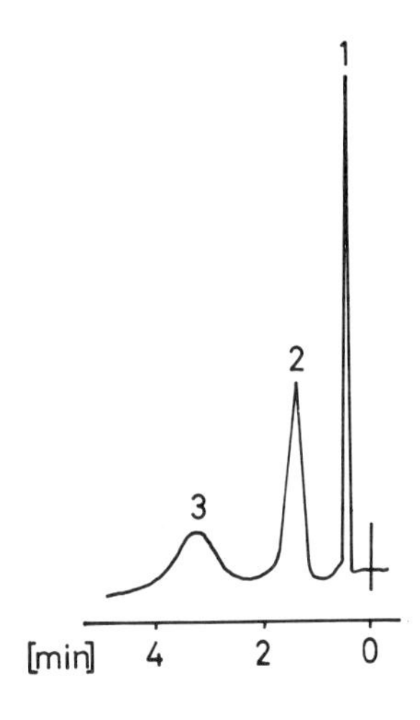

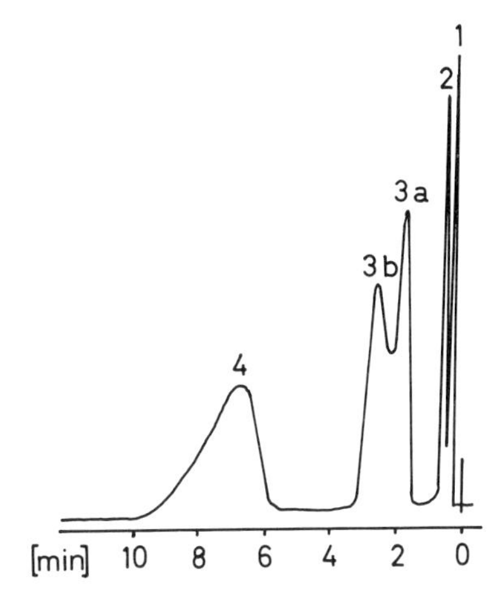

Abb.VIII.7. Separation of vitamins. Ion exchanger: n-butylsulfonic acid on silica Si 100; d_p ~ 10 µm; 230 µeq/g; column: 50 cm, 2.3 mm i.d.; eluent: 0.02 M sodium phosphate buffer; left: pH = 5.5; u = 2.2 cm/sec; Δp = 90 atm; 1 = thiamine 2HCl; 2 = nicotinic acid amide; 3 = pyridoxyl HCl; right: pH = 3.9; u = 3.6 cm/sec; Δp = 150 atm; 1 = ascorbic acid; 2 = nicotinic acid; 3 = nicotinic acid amide; 4 = pyridoxyl HCl

3. Ion Exchangers on Chemically Modified Silica Gel

These brush-type ion exchangers combine good chromatographic properties with a high exchange capacity [11-13,18]. Both amino and nucleic acid analyses [11], as well as the separation of water-soluble vitamins, have been effected on such materials. Fig.VIII.7 shows the separation of water-soluble vitamins at various pH values on a cation exchanger [18]. At a pH of 3.9 nicotinamide appears as a double peak, but on raising the pH to 5.5 it elutes as a single peak. At the same time, the analysis time is shortened considerably.

However, even such ion exchangers exhibit "reversed-phase" sorption on the matrix [18]. A review of the latest applications of ion exchange chromatography may be found in *Analytical Chemistry* [49].

References Chapter VIII

1. Samuelson, O.: Ion Exchange Separations in Analytical Chemistry. New York: Wiley 1963
2. Helfferich, F.: Ionenaustauscher. Weinheim: Verlag Chemie 1959
3. Dorfner, K.: Ionenaustauscher. 2. Aufl. Berlin: De Gruyter 1964
4. Inczêdy, J.: Analytische Anwendungen von Ionenaustauschern. Budapest: Verlag der Ungar. Akademie der Wissenschaften 1964
5. Riemann III, W., Walton, H.F.: Ion-exchange in Analytical Chemistry. Oxford: Pergamon Press
6. Hesse, G.: Chromatographisches Praktikum. Frankfurt: Akadem. Verlagsges. 1968
7. Davankov, V.A., Semechkin, A.V.: J. Chromatogr. *141*, 313 (1977)
8. Martinsson, E., Samuelson, O.: J. Chromatogr. *50*, 429 (1970)
9. Horvath, C., Preiss, B., Lipsky, S.R.: Anal. Chem. *39*, 1422 (1967)
10. Kirkland, J.J.: J. Chromatogr. Sci. *8*, 72 (1970)
11. Unger, K., Nyamah, D.: Chromatographia *7*, 63 (1974)
12. Weigand, N., Sebestian, I., Halâsz, I.: J. Chromatogr. *102*, 333 (1975)
13. Saunders, D.H., Barford, R.A., Magidam, P., Olszewski, L.T., Rothbart, H.L.: Anal. Chem. *46*, 834 (1974)
14. Horwitz, E.P., Delphin, W.H., Bloomquist, C.A.A., Vandegrift, G.F.: J. Chromatogr. *125*, 203 (1976)
15. Eksborg, S., Schill, G.: Anal. Chem. *45*, 2092 (1973)
16. Kraak, J.C., Huber, J.F.K.: J. Chromatogr. *102*, 333 (1975)
17. Eksborg, S., Lagerström, P.O., Modin, R., Schill, G.: J. Chromatogr. *83*, 99 (1973)
18. Weigand, N.: Dissertation Saarbrücken 1974
19. Knox, J.H., Jurand, J.: J. Chromatogr. *87*, 95 (1973)
20. Khym, J.X., Zill, L.P.: J. Am. Chem. Soc. *74*, 2090 (1952)
21. Bauer, H., Voelter, W.: Chromatographia *9*, 433 (1976)
22. Benson, J.V., Gordon, M.J., Patterson, J.A.: Anal. Biochem. *18*, 228 (1967)
23. Ertinghausen, G., Adler, H.J., Reichler, A.S.: J. Chromatogr. *42*, 355 (1969)
24. Hamilton: Application sheets
25. Udenfried, S., Stein, S., Bohlen, P., Leimgruber, W., Weigele, M.: Science *178*, 871 (1972)
26. Benson, J.R.: Durrum Resin Report No. 6, December 1973
27. Scott, C.D., in: Kirkland, J.J.: Practice of Modern Liquid Chromatography. New York: Wiley-Interscience 1971
28. Scott, C.D.: Science *186*, 226 (1974)
29. Gere, D.R., in: Kirkland, J.J.: Modern Practice of Liquid Chromatography. New York: Wiley-Interscience 1971
30. Brown, P.R.: High Pressure Liquid Chromatography, Biochemical and Biomedical Application. Academic Press 1973
31. Horvath, C.S., in: Glick, F. (Ed.): Methods of Biochemical Analysis. New York: Wiley 1973

32. Horvath, C.S., in: Marinsky, J.A., Marcus, Y. (Eds.): Ion Exchange and Solvent Extraction, Vol. 5. New York: Dekker 1973

33. Scott, C.D., Chilcote, D.D., Lu, N.E.: Anal. Chem. *44*, 85 (1972)

34. Scott, C.D., Jolley, R.L., Pitt, W.W., Johnson, W.F.: Am. J. Clin. Pathol. *53*, 701 (1970)

35. Scott, C.D., Lee, N.E.: J. Chromatogr. *83*, 383 (1973)

36. Burtis, C.A.: J. Chromatogr. *52*, 97 (1970)

37. Scott, C.D., in: Bodansky, O., Latner, A.L. (Eds.): Advances in Clinical Chemistry, Vol. 15. New York: Academic Press 1972

38. Liljamaa, J.J., Hallén, A.A.: J. Chromatogr. *57*, 153 (1971)

39. Hobbs, J.S., Lawrence, J.G.: J. Chromatogr. *72*, 311 (1972)

40. Application sheet 05/05 Fa. Siemens AG, Karlsruhe, G.F.R.

41. Kaiser, U.J.: Chromatographia *6*, 387 (1973)

42. Lange, H.W., Hempel, K.: J. Chromatogr. *59*, 53 (1971)

43. Brown, P.R.: J. Chromatogr. *57*, 383 (1971)

44. Shmukler, H.W.: J. Chromatogr. Sci. *10*, 137 (1972)

45. Schmit, J.A., in: Kirkland, J.J.: Modern Practice of Liquid Chromatography. New York: Wiley-Interscience 1971

46. Anders, M.W., Latorre, J.P.: J. Anal. Chem. *42*, 1430 (1970)

47. Anders, M.W., Latorre, J.P.: J. Chromatogr. *55*, 409 (1971)

48. Henry, R.A., Schmit, J.A., Williams, R.C.: J. Chromatogr. Sci. *11*, 358 (1973)

49. Walton, H.F.: Anal. Chem. *46*, 398 R (1974)

50. Eksteen, R., Kraak, J.C., Linssen, P.: J. Chromatogr. *148*, 413 (1978)

Chapter IX

Exclusion Chromatography Gel Permeation Chromatography

A. Introduction

In contrast to the separation techniques discussed up to now, separation by exclusion chromatography involves a single, unambiguous mechanism. In the absence of interactions (by definition) between the sample and the stationary phase surface, the elution order (or elution volume) is solely a function of the molecular size (see IX.B). Hence, this method is eminently suited for the separation of high-molecular-weight (polymeric) samples.

This technique is known by several names. Initially, only gels (= eluent-swollen polymers) with various degrees of cross-linking were used, which gave rise to designations such as gel filtration [1], gel permeation chromatography, or gel chromatography. However, because the separation mechanism is based on the fact that access to the pores is excluded for certain molecules, and since gels are unsuitable for HPLC because of their low pressure stability, the designation of exclusion chromatography (EC) is preferred here. Both hydrophylic (e.g., cross-linked dextrans) [1] and hydrophobic materials (e.g., cross-linked polystyrenes) [2] have been used as stationary phases. Rigid materials such as silica, porous glasses, etc. having a definite pore size distribution also can be used for EC if interactions of the samples with the surface are excluded by proper eluent selection. Several monographs [3-6] are devoted to the classical column-chromatographic procedures. Here only the HPLC applications of exclusion chromatography will be described.

B. Basics of Exclusion Chromatography

Porous solids can be characterized by their specific surface area, their pore volume and their pore size distribution. The pore volume of a solid is accessible only to molecules whose largest diameter is smaller than the pore opening. Molecules that can diffuse into the pore volume, where no transport occurs (stagnant mobile phase), migrate more slowly through the column than those that are excluded. Because the pore size distribution is never strictly monodisperse but varies over a certain range, a sharp division between excluded and permeating molecules obtains only if the sample has an uniform molecular weight. If this is not the case, the following occurs:

All molecules that are too large to diffuse into the pores elute at the retention volume equal to the interstitial volume, i.e., that between the particles. The column then behaves as if it were packed with non-porous inert glass beads (cf. II.B). Molecules with diameters smaller than the pore opening penetrate to various extents and are thus retarded relative to those excluded. This retardation (larger elution volume) increases with deeper penetration into the pores, i.e., with greater accessibility of the pore volume. Since by definition there is no interaction between the sample and the stationary phase surface in exclusion chromatography, the elution of all molecules separated on the basis of size is not complete until the smallest one, which has access to all the pores, reaches the end of the column. This elution volume is defined as the column dead volume V_o and corresponds to the dead time t_o. All sample components that emerge after this dead time are additionally retarded by interacting in some way with the stationary phase. The dead volume V_o consists of the interstitial volume V_z and the pore volume V_p. The difference in the retention volumes between the totally excluded and the smallest molecules corresponds to the pore volume of the column packing.

Fig.IX.1 presents these relationships schematically. The upper portion contains the elution curves of individual polymer standards, whereas the lower shows the relationship between elution volume and polymer size. Such "calibration curves" may serve to draw inferences concerning the molecular weight distribution. Of course, the molecular weight is only a measure of the coil diameter of the polymer molecule in the eluent.

It is impractical to employ the usual chromatographic quantities such as the retention time, k' value, relative retention, etc. The smallest molecules always possess the longest retention time. To avoid errors, only the elution volume V_e will be used here. The elution volume is a function of the molecular diameter of the sample and corresponds to the sum of the interstitial volume V_z and the pore volume V_p accessible to the sample. Totally excluded molecules elute with

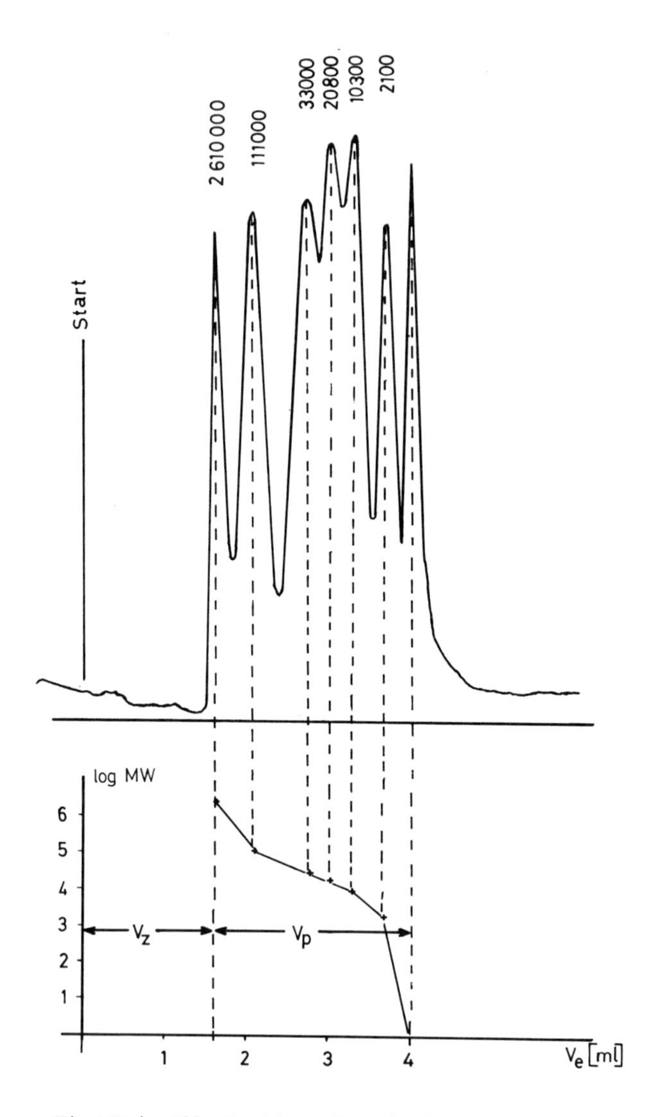

Fig.IX.1. Illustration of exclusion chromatography. Upper portion: Separation of polystyrenes with molecular weights from 2100 to 2.6 million (benzene was used as the smallest molecule). Lower portion: The calibration curve derived for exclusion chromatography; V_z = interstitial volume; V_p = pore volume of the stationary phase. Experimental conditions: silica gel; $d_p \sim 10$ µm; V_p = 2.05 ml/g; column: 30 cm, 4 mm i.d.; u = 0.12 cm/sec; Δp = 22 atm; eluent: methylene chloride

$V_e = V_z$, whereas the smallest (e.g., solvent molecules) have $V_e = V_z + V_p = V_o$ (as illustrated schematically in Fig.IX.1).

This statement can be made more general if this volume is replaced by a quantity based on the empty column volume (V_k) - the porosity (cf. Chapter II.B) [7], which can be determined easily and is independent of the column used (the packing density, etc.). The dead volume is related to the total porosity ε_T which represents the fraction of the column volume available to the stagnant and moving eluent, that is $\varepsilon_T = V_o/V_k$. The volume of a column may be determined with any desired accuracy by weighing it empty and filled with water. For exclusion-chromatographic separations only the porosity ε_p is used, which is proportional to the pore volume (V_p) of the support ($\varepsilon_p = V_p/V_k$). The larger the molecules, the smaller is the accessible portion of the porosity and the shorter are the retention times.

An enhancement of the resolution requires the largest possible pore volume, in addition to a certain number of theoretical plates. In this respect, a large available pore volume in the column (cm^3/cm^3) is much more important than a large specific pore volume (cm^3/g). In exclusion chromatography large differences in the specific pore volume do not have the effect on resolution, expected at first glance, because the apparent density of the material changes as well. Thus, doubling the specific pore volume from 0.5 cm^3/g to 1 cm^3/g raises the effective pore volume within the column by only about 30% [18].

Besides pore volume, column efficiency also plays an important role. Due to the lower diffusion coefficients of polymer samples, the H values in GPC are always greater than those observed for low-molecular-weight compounds on the same column. Moreover, individual compounds are never involved but rather polymer fractions which, no matter how narrow, still have a finite distribution. Further separation of the fractions on a column yields misleadingly high H values. The highest H values are obtained when the maximum of the pore size distribution and the coil diameter of the polymer fraction coincide closely [19]. Therefore, column efficiency should only be based on the H values of fully excluded samples or, even better, on that of the smallest molecule, which usually is a single substance with a definite molecular weight.

The effective pore volume as well as the efficiency can be increased by connecting several columns in series. The same effect can be achieved by recycling the sample through the same column. However, recycling is practical only if the band spreading in the connections and in the pump is small.

Extra-column band broadening may substantially reduce the column efficiency. This phenomenon increases the H values of all substances and becomes most pronounced with short columns, especially with $L < 15$ cm, due to a small V_k. For example, an increase in the column diameter from 3.2 mm to about 8 mm for a length of 10 cm (a *ca.* 6-fold volume increase) raised the efficiency from 1500 to 5000 plates ($d_p \sim 6$ µm) [18]. This illustrates the effect of extra-column band broadening.

C. Stationary Phases for Exclusion Chromatography

Stationary phases for exclusion chromatography have already been summarized in Section V.D. Most of the stationary phases from classical GPC are unsuitable because of their compressibility. Some highly cross-linked polystyrene gels (e.g., µ-Styragel®) and poly(acrylate-ethylene glycol) gels can be employed at pressures up to 50 atm.

Pressure-stable solids with a rigid matrix, such as silica gel or porous glass, can be used for exclusion chromatography under HPLC conditions. These materials possess certain advantages over the soft gels: They are easier to pack, do not need to be pre-swollen in the eluent, and yield mechanically stable columns (i.e., the permeability is not a function of the applied pressure). For practical application many more eluents can be used, as there is no need to assure adequate swelling of the gels. The eluent can be changed easily because the degree of swelling does not change. This enhances the advantages and versatility of the method. An additional advantage lies in the resistance of these solids and their pore structure to virtually all organic solvents, even at elevated temperatures. This is important, for instance, in the characterization of poly-olefins.

Silica gels can be produced with pore diameters ranging between 20 Å and 25000 Å. Spherical silica gels with d_p around 10 µm or less are available with average pore diameters from 60 to 4000 Å, and can be used to separate most ordinary polymeric molecules. Species with a molecular weight < 1000 can be separated on silica gel having a pore size of 60 Å, whereas even polymer standards with a molecular weight of $7 \cdot 10^6$ are not fully excluded when the pore size is 4000 Å [18]. A silica gel with an average pore diameter of about 250 Å will separate polystyrene samples with molecular weight ranges from 2000 to about 100 000 (cf. Fig.IX.1).

The disadvantage of these polar stationary phases stems from their adsorptive properties. In many cases, however, this activity can be eliminated by a judicious selection of the eluent (cf. the Eluotropic series, Table VI.2). Polystyrenes are adsorbed on silica gel from carbon tetrachloride ($V_e > V_o$), but are excluded from methylene chloride, tetrahydrofuran, and dimethylformamide, i.e., they are separated on basis of their molecular size ($V_e < V_o$). Any interfering residual activity can be eliminated by silanizing the surface with trimethylchlorosilane. If more carbon is bonded to the surface, as in reversed phases, the pore volume decreases in proportion to the amount of carbon bonded.

However, this type of stationary phase still retains sufficient catalytic activity to alter or adsorb irreversibly many natural polymers (proteins, enzymes, etc.) [13]. Such phases are unsuited for aqueous systems because they operate on the same principle as reversed phases (cf. Section VI.II) and adsorb the samples, thereby preventing separations based solely on molecular size. By using organic residues with polar functional groups, chemically bonded phases can be prepared that, like silica gel, are wetted by water. A bonded phase with the following functional group

$$Si-CH_2-CH_2-CH_2-O-CH_2-CHOH-CH_2OH$$

is frequently employed [20,21], and is commercially available, for example, under the name of Glycophase or DIOL. Most important is the presence of the epoxide group that can be reacted with water (to give the glycol) and other nucleophiles such as amines, alcohols, etc. [21]. But even these phases adsorb some proteins and enzymes irreversibly [20]. Other phases (i.e., bonded amides) have been discussed and employed for the aqueous exclusion chromatography of proteins [25].

Stationary phases are characterized by their exclusion limits and by calibration curves. The latter consist of plots of the logarithm of the weight-average molecular weight $\bar{M}_w$ against the elution volume V_e (cf. Fig.IX.1). These curves always have a linearly rising portion where $\log M_w = f(V_e)$, which corresponds to the optimal area of application of each phase. Such characterization curves, however, do depend on the polymer samples used. This has led to numerous attempts to establish universal calibration curves, and "rotation coil diameters" have been assigned to polymers as a function of their average molecular weight. In addition, many other variously defined radii have been used to characterize supports [6,8-10].

The pore size distribution of organic polymer gels differs from that of the silica gels [26]. The organic gels have a relatively broad pore size distribution, which begins with small diameters, and terminates with a very sharply defined upper exclusion limit. The silica gels have a more or less narrow pore size distribution disposed symmetrically about the mean pore diameter. To be able to cover the entire range of soft gels with silica, several columns containing silicas having different pore size distributions must be connected in series. It was shown [27] to be more than adequate to utilize only two silicas, whose average pore sizes differ by a factor of about 10. Thus, by connecting two columns packed with 60 Å and 500 Å pore diameter materials, polystyrenes with molecular weights from 210 (dimer) to 2.6 million could be separated (Fig.IX.2). This column combination (30 cm of Si 60 and 40 cm of Si 500) generated 30 000 plates based on the unretained peak (benzene).

D. Applications of Exclusion Chromatography

The principles of column chromatography are also valid for exclusion chromatography. Doubling the column length doubles the column volume. Long columns are preferred because they provide a more satisfactory pore volume and separation efficiency. As a corollary, band spreading should be kept to a minimum. In this respect, small-diameter particles are particularly efficient. The eluent viscosity should be minimized in order to enhance the rate of diffusion. This is one reason (the other being increased polymer solubility) why exclusion chromatographic separations, in particular, are frequently performed at higher temperatures, sometimes just below the boiling point of the eluent. The appearance of gas bubbles in the detector cell may be prevented by means of a low back pressure (1 - 2 atm).

The discussion of special applications is beyond the scope of this book. Examples of applications may be found in the monographs mentioned [3-6]. Only the potential of rapid exclusion chromatography will be illustrated. Frequently, for the separation of very complex mixtures, exclusion chromatography furnishes a method for the preliminary removal of high molecular weight impurities.

It should be pointed out that risks may be involved in using the high linear velocities attainable in HPLC with rigid gels. It was

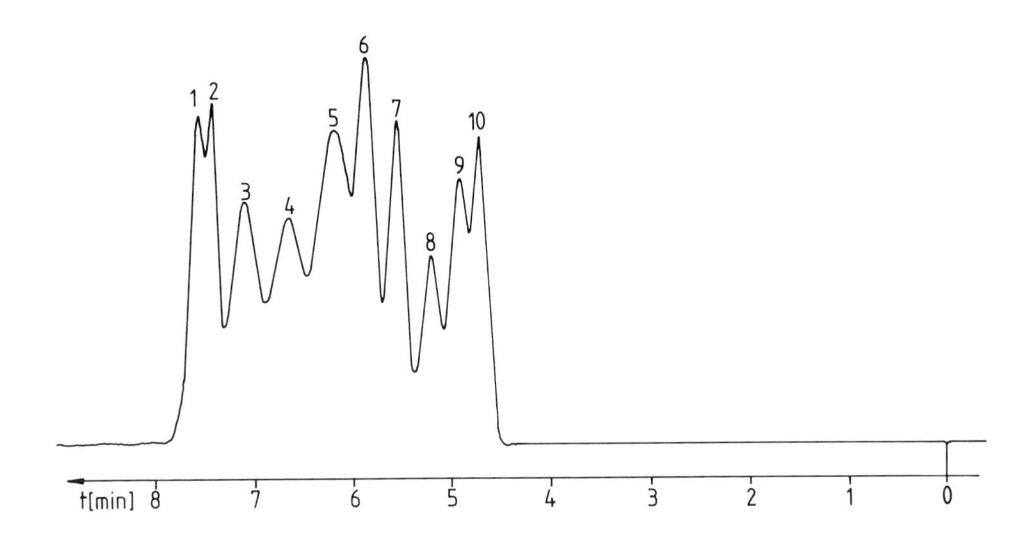

Fig.IX.2. High efficiency gel permeation. Column: 30 cm silica Si 60 + 40 cm silica Si 500, 4.2 mm i.d.; d_p ~ 10 µm; eluent: methylene chloride; F = 1 ml/min. Samples: 1 = benzene (~ 30 000 plates); 2 = distyrene; 3 - 10 = polystyrene with $\bar{M}_w$ 800, 9000, 20 800, 50 000, 110 000, 233 000, 2 610 000

reported [18] that very large molecules (e.g., with M_w of $7 \cdot 10^6$) may be cleaved by the shear forces at linear velocities > 1 mm/sec.

Recently, the application of EC to small molecules (M_w < 600) was demonstrated with the separation of aliphatic hydrocarbons on a narrow-pore gel (exclusion limit 40 Å) with THF as eluent [28] (Fig. IX.3). It is evident that the separation efficiency decreases with increasing molecular weight. This contrasts with chromatography on nonpolar phases where the relative retentions of the members of a homologous series remains constant. This separation, therefore, could probably have been performed more propitiously on an RP. However in that case, the retention increases with increasing molecular weight, and it is doubtful whether the separation of the C_6 - C_{32} could have been effected on an RP under isocratic conditions. Gradient elution could not be applied because of the RI detector.

1. Determination of the Molecular Weight Distribution of Polymers

The primary area of application of exclusion chromatography remains the determination of the molecular weight distribution of polymers.

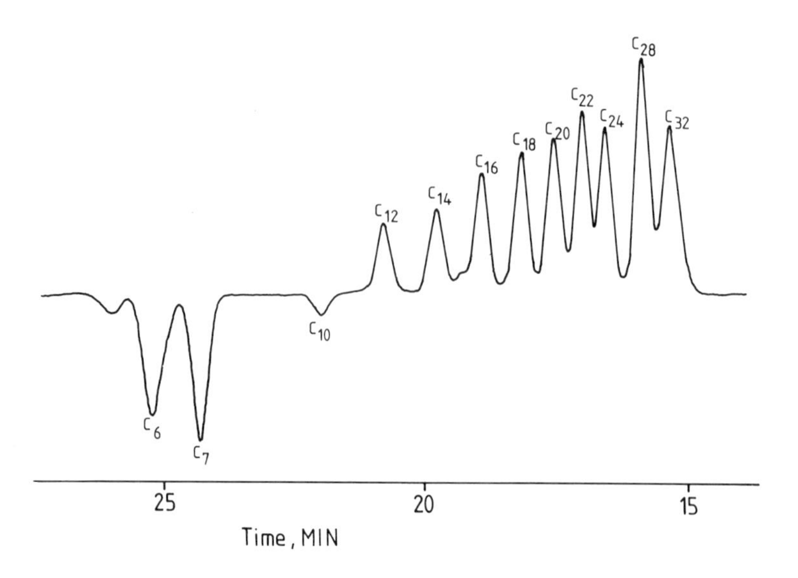

Fig.IX.3. Separation of aliphatic hydrocarbons C_6 - C_{32}. Column: Micropak TSK Gel, 40 A x 2; eluent: tetrahydrofuran; F = 1 ml/min; (courtesy R. Majors, Varian Assoc.)

It is assumed that the sample molecules are not adsorbed on the support surface and that the "distribution equilibrium" [11] of the sample between the stagnant and moving eluent is established instantaneously.

A calibration curve should be constructed as illustrated in Fig.IX.1, using the appropriate polymer standards, such as polystyrenes with a narrow molecular weight distribution. Good agreement [8] is obtained between chemically dissimilar linear polymers by utilizing the hydrodynamic radii of the molecules, which are proportional to the logarithm of the molecular weight and the intrinsic viscosity (Staudinger Index) $[\eta]$. The values determined in this way have an uncertainty of about 5% for a molecular weight of 50 000 [12]. Sometimes the results are less satisfactory because the polymers are not always present in the form assumed by the simple model described here. However, a qualitative picture of the composition or of the molecular weight distribution of the polymer sample is obtained in all cases. A calibration curve for a certain polymer should be verified by an independent molecular weight determination. After changing the column or, especially, the batch of the support, it is essential to recalibrate with the same standards, inasmuch as it is nearly impossible to obtain two batches of silica with identical pore size distributions. Since a change of eluent does not alter the pore structure of the rigid silica gels, there is no change in the selectivity of the separation system, provided the configuration of the dissolved polymer molecule is independent of the type of solvent used.

The separation of polymers rarely yields chromatograms like those shown in Figs. IX.1 and IX.2, which were obtained with relatively narrow standards (polydispersity < 1.1). More commonly, a single more or less broad peak is obtained. Polymers are generally characterized in terms of their molecular weight distribution or polydispersity, i.e., the ratio of weight-averaged to number-averaged molecular weights. The width of the eluted peak is only to a first approximation proportional to the molecular weight distribution, because it is distorted by diffusion processes and as a result of instrumental band broadening. Furthermore, the detector response is not independent of the molecular weight. For the determination of the actual MW distribution from an elution diagram laborious corrections have been described to eliminate the effect of band broadening (see the discussion in [4]).

Various methods are available for obtaining a molecular weight distribution from an elution diagram [4]. Chromatograms containing separate or only partially overlapping peaks, which are often observed in the separation of monomers and oligomers, can be interpreted in the usual manner: After identification of the peaks, the area under each is measured to give the relative concentration, and the molecular weight distribution can be calculated.

For high polymer samples, where only a single peak is obtained, this approach is no longer possible. To determine the polydispersity the peak must first be divided into separate equal segments along the baseline. To cover an extensive range of molecular weights a long linear section in the calibration curve (log MW *vs.* V_e) is desirable. A representative molecular weight based on the calibration curve for the column can be assigned to the average elution volume of each segment. Usually, this is the weight-averaged MW ($\bar{M}_w$) of the calibration standards. The $\bar{M}_w$ of the sample is calculated as the sum of the products of the average peak height h of each segment and the corresponding molecular weight, divided by the sum of the heights h_i of the arbitrarily assigned segments:

$$\bar{M}_w = \frac{\sum^{i} h_i M_{wi}}{\sum^{i} h_i} .$$

The number-averaged molecular weight $\bar{M}_N$ is calculated similarly:

$$\bar{M}_N = \frac{\sum^i h_i}{\sum^i \frac{h_i}{M_{wi}}} \ .$$

These calculations, however, are not corrected for diffusion or for the varying relative detector response.

Corrections for diffusion have been described [4] that essentially eliminate the band broadening contribution and usually make the peak narrower. The correction for relative response is more tedious and involves the collection of narrow fractions, rechromatographing known amounts, and obtaining a calibration by measuring the areas.

The average molecular weights obtained by this relatively simple procedure (see Waters GPC manual) need not necessarily agree with those derived from classical methods such as osmometry, viscosity, etc. Laborious procedures exist [4] for improving the agreement between the values obtained by these methods, but their discussion is beyond the scope of this text. A decisive advance, by molecular weight determination directly in the column effluent, could be achieved with a light scattering detector.

In addition to the molecular weight distribution of polymers, the particle radii of polymer dispersions have also been determined by EC [10]. Dispersions of polymethylmethacrylate having particle diameters of 350 to 2390 Å could be separated on a silica gel whose average pore diameter was 12 000 Å. The eluent was water, sometimes with the addition of an emulsifier. Because of the low "rate of diffusion" of the particles, only slow eluent velocities (~ 0.1 ml/min for a 9 mm i.d. column) could be used. The elution curves for the various dispersions (350 - 2390 Å) are shown in Fig.IX.4. The calibration curves for polystyrene and methylmethacrylate dispersions are identical. A filtration effect was observed for ~ 1 μm diameter particles, which were subsequently recovered by backflushing the column.

By reversing this calibration procedure, a means for the rapid determination of the pore size distribution of solids is obtained [7]. The polystyrene standards are arbitrarily assigned exclusion values (in Å) that make the pore size distribution curves measured by EC correspond with those determined by classical procedures (B.E.T., mercury porosimetry). On the basis of these measurements, each of the poly-

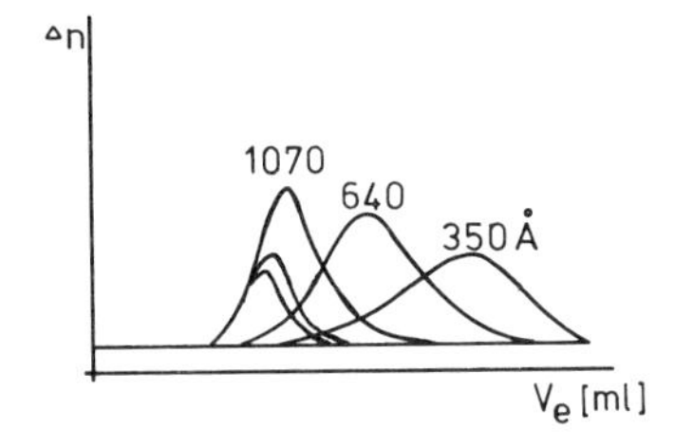

Fig.IX.4. Exclusion chromatography of polymer dispersions [10] (poly(methylmethacrylate)). Stationary phase: Merckogel Si 3500; eluent: water; column: 160 cm, 9 mm i.d.; F = 8 ml/h. The 1070 Å particle peak also contains the larger particles

styrene standards is assigned a pore diameter ∅ at which it is excluded. The relationship of this ∅ value to the molecular weight is given as follows:

$$\varnothing = 0.62\ (\bar{M}_w)^{0.59}$$

where $\bar{M}_w$ is the weight-averaged molecular weight. Accordingly, a polystyrene sample (dissolved in methylene chloride) with $\bar{M}_w$ = 10 000 and its coil diameter correspond to a pore diameter of 140 Å; a molecular weight of 3.7 million corresponds to a pore diameter of 4530 Å. Surprisingly, the pore diameter assigned to the polymers is found to be 2.5 times [14] that of the coil diameter determined by other means [15], i.e., the ∅ of the solid must be 2.5 times the coil diameter of the polymer in order to establish an instantaneous distribution equilibrium.

This approach for the determination of pore size distribution curves can be applied to fine powders (d_p > 1 μm) as well as to coarse particles (d_p < 150 μm). The upper limit of the measurable pore diameters is currently restricted by the available polymer fractions. The pore size distribution of eluent-swollen gels [26] can also be determined by this method.

2. Application of Rapid Exclusion Chromatography to Biochemical Problems

The use of pressure-stable silica gels for the EC determination of proteins is restricted because of irreversible sorption [13]. This effect could be reduced, or even completely eliminated in some cases, by using a porous glass whose surface was chemically bonded with "carbohydrates". Fig.IX.5 shows the separation of *human serum* on a chemically modified porous glass. Such separation would be impossible on bare

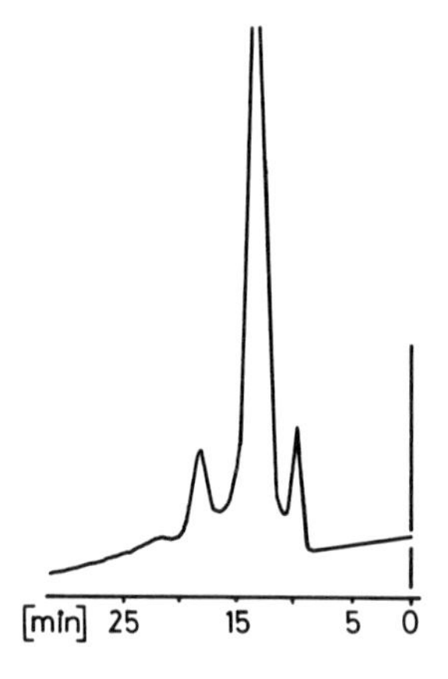

Fig.IX.5. EC separation of human serum. Stationary phase: Glycophase-G on CPG (controlled pore glass). 170 Å GTS; column: 100 cm, 4.2 mm i.d.; eluent: 0.05 M phosphate buffer (pH 7.0). The principal peak is attributed to albumin (MW 70 000) (Pierce Previews, June 1974)

porous glass due to irreversible adsorption. Unfortunately, some proteins (hemoglobin, catalase, etc.) are still held, in part irreversibly, even on such modified stationary phases. Fig.IX.6 shows an elution profile of dextrans and polyethylene glycol 600 on such a glycolphase. Whereas the dextrans elute before the unretained substance, D_2O, and are resolved on the basis of their molecular sizes, the polyethylene glycol is adsorbed from H_2O (the eluent), and partially separated into homologous polymeric members. Higher molecular weight polyethylene glycols are even more strongly retarded under these conditions. The sorption of the polyethylene glycols can be repressed by changing the eluent composition or the polar, bonded functional groups, so that they too can be separated by EC.

Water-soluble polymers such as adhesives and polyvinyl alcohols [22], as well as dextrans and chitosan (deacetylated chitin) [23], and other water-soluble natural products or biological extracts [20, 21,24] have been resolved on glycol phases. For the latter materials especially, complete elimination of the sorption or catalytic activity on the stationary phase has not yet been achieved.

Chemically bonded phases with other hydrophilic functional groups have been investigated for their suitability for the EC of water-soluble polymers [25]. The more hydrophilic the bonded organic moiety,

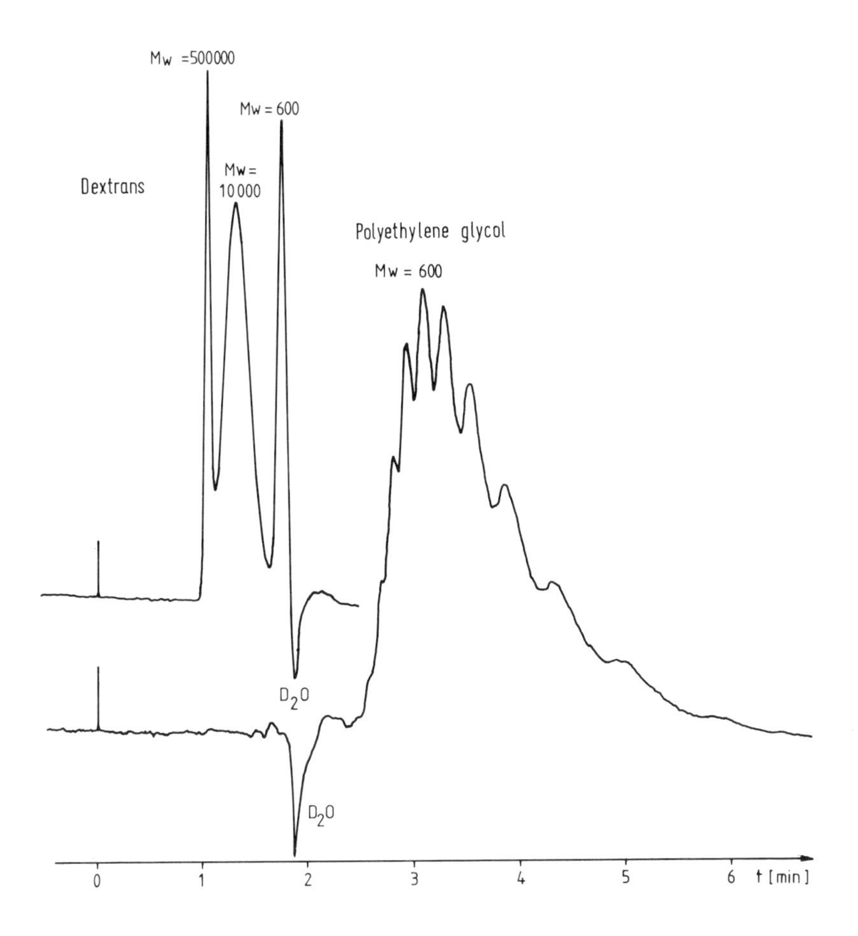

Fig.IX.6. Glycol phase. Elution of dextrans and polyethylene glycol. Stationary phase: Si-CH_2-CH_2-CH_2-O-CH_2-CHOH-CH_2OH on Si 100; d_p ~ 10 µm; L = 30 cm, 4.3 mm i.d.; eluent: water; F = 2.1 cm^3/min; u = 2.6 mm/sec; Δp = 67 atm; dextrans: M_w 500 000, M_w 10 000; raffinose (M_w 595); polyethylene glycol M_w 600; unretained substance D_2O; differential refractometer

the shorter are the retention times of the polyethylene glycols, for example. On diamine, amide, or glycinamide phases, these polymers are no longer sorbed but separated exclusively by an EC mechanism. On an amide phase proteins and enzymes could be eluted quantitatively with an aqueous buffer, the enzymes with full retention of activity. The elution volumes of the proteins could be correlated with those of the polystyrenes because the pore volumes of such rigid phases are independent of the eluent. This was possible, however, only after reducing the exclusion effects, due to the Donnan potential, as well as sorption

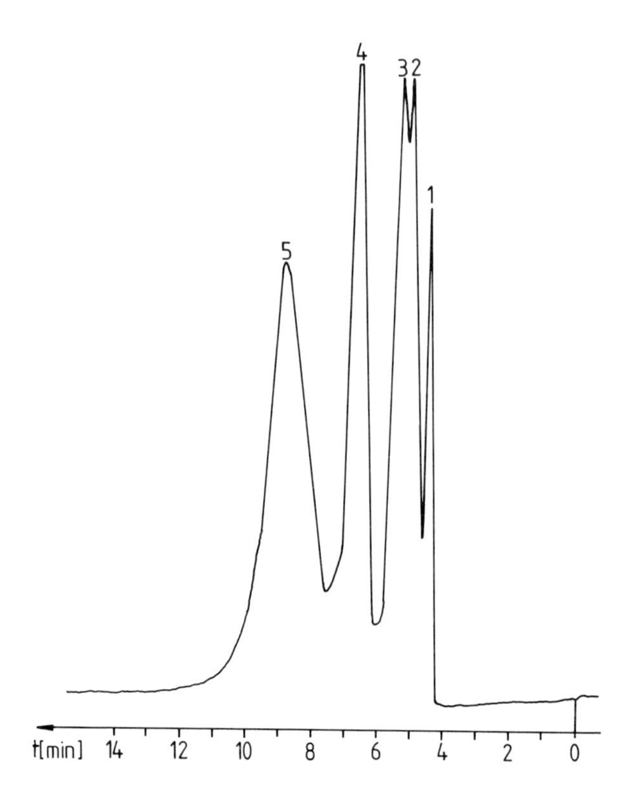

Fig.IX.7. Separation of proteins. Stationary phase: Si-CH_2-CH_2-CH_2-NH-CO-CH_3 on silica Si 100; d_p ~ 10 µm; L = 30 cm, 4.2 mm i.d.; eluent: 0.1 m Tris-buffer, pH 7.5 + NaCl to ionic strength 0.5; F = 0.9 ml/min; p = 60 atm; 1 = ferritin; 2 = BSA; 3 = ovalbumin; 4 = myoglobin; 5 = chymotrypsinogen A

effects, by raising the ionic strength of the buffer. If the ionic strength of the eluent was > 0.5, the protein elution volumes were solely a function of the molecular size and the pore size distribution of the silica support.

Fig.IX.7 shows the EC of proteins on a chemically bonded silica having an average pore size of 100 Å [29]. To separate the albumins or larger proteins optimally, a silica with a pore size of 200 - 350 Å would be required.

Some hydrophilic organic polymers, such as a copolymer of ethylene glycol dimethacrylate [16], are relatively pressure-stable. Such products, e.g., ®Merckogel PGM 2000, are supposed to be usable at pressures of up to 100 atm. Unfortunately, it has not yet been possible to produce these gels with various exclusion limits. However, they can be used for the EC separations of unstable substances [17].

References Chapter IX

1. Porath, J., Flodin, P.: Nature *183*, 1657 (1959)
2. Vaughan, M.F.: Nature *188*, 55 (1960)
3. Determann, H.: Gel-Chromatographie. Berlin-Heidelberg-New York: Springer 1967
4. Altgelt, K.H.: Advanc. Chromatography *7*, 3 (1968)
5. Bombaugh, K.J., in: Kirkland, J.J. (Ed.): Modern Practice of Liquid Chromatography. New York: Wiley 1971
6. Altgelt, K.H., Segal, L. (Eds.): Gel Permeation Chromatography. New York: Dekker 1971
7. Halâsz, I., Martin, K.: Ber. Bunsenges. *79*, 731 (1975)
8. Benoit, H., Gallot, Z., in: Kovats, E. (Ed.): Säulenchromatographie 1969. Supplementum zu Chimia. Aarau 1970
9. Kreveld, M.E., v. Denhoed, N.: J. Chromatogr. *83*, 111 (1973)
10. Krebs, K.-F., Eisenbeiß, F.: Vortrag GDCh-Hauptversammlung 1971; cf. Eisenbeiß, F.: Kontakte *3*, 35 (1973)
11. Casassa, E.F., Tagami, Y.: Macromolecules *2*, 19 (1969)
12. Cooper, A.R., Johnson, J.F., Porter, R.S.: Int. Lab. May., June 1973, p. 38
13. Kennedy, J.F.: J. Chromatogr. *69*, 325 (1972)
14. Martin, K., Halâsz, I.: Angew. Chem. to be published (1978)
15. Vollmert, B.: Polymer Chemistry. Berlin-Heidelberg-New York: Springer 1974
16. Heitz, W., Winan, H.: Makromolekulare Chem. *131*, 75 (1970)
17. Peters, R.: Kontakte *3*, 22 (1973)
18. Kirkland, J.J.: J. Chromatogr. *125*, 231 (1976)
19. Werner, W.: Dissertation Saarbrücken 1976
20. Regnier, F.E., Noel, R.: J. Chromatogr. Sci. *14*, 316 (1976)
21. Chang, S.H., Gooding, K.M., Regnier, F.E.: J. Chromatogr. *120*, 321 (1976)
22. Persiani, C., Cuker, P., French, K.: J. Chromatogr. Sci. *14*, 417 (1976)
23. Wu, A.C.M., Bough, W.A., Conrad, E.C., Alden jr., K.E.: J. Chromatogr. *128*, 87 (1976)
24. Chang, S.M., Gooding, K.M., Regnier, F.E.: J. Chromatogr. *125*, 103 (1976)
25. Engelhardt, H., Mathes, D.: J. Chromatogr. *142*, 311 (1977)
26. Vogtel, P.: Dissertation Saarbrücken 1977
27. Yau, W.W., Ginnard, C.R., Kirkland, J.J.: J. Chromatogr. *149*, 465 (1978)
28. Majors, R.E.: Varian Assoc., private communication
29. Mathes, D.: Dissertation Saarbrücken 1978

Chapter X

Selection of the Separation System

The beginner is frequently confronted with the problem of selecting the separation system that will provide most rapidly an optimal resolution. This problem becomes particularly difficult when nothing is known about the properties or composition of the sample.

A preliminary decision can be based on sample solubility in the standard HPLC solvents, as outlined in Fig.X.1. Among the proven "non-polar" eluents are n-heptane (or pentane to isooctane), 1-chloropropane, methylene chloride (or chloroform), perhaps with the addition of 5 - 10% ethyl acetate. Mixtures may also be used, of course. The "polar" include water, methanol, other lower alcohols, and acetonitrile.

Solubility in, or miscibility with, one of the polar eluents indicates the presence of highly polar components, whose resolution by adsorption chromatography on silica or alumina is unlikely to be successful. Elution with very polar eluents (e.g., alcohols, acetonitrile) might be possible, but experience shows such separations to be poor. Instead, such samples should be more amenable to separation on a non-polar stationary phase system (cf. VI.II) with a polar eluent such as water, methanol or its mixtures. A partition system, e.g., one involving a ternary mixture, may be suitable for separating and eluting the sample components. For salt-like compounds it would be logical to attempt an ion-exchange separation.

There are, of course, separation problems that can be solved more or less equally well with any system. Thus, steroids may be separated by adsorption (cf. Fig.VI.4), on a reversed-phase system (cf. Fig.VI.26) or by partition (cf. Fig.VII.8). In these systems the retention may be varied by minor modifications of the eluent composition. In deciding ultimately which system to use, knowledge of and experience with a particular separation system are important factors to consider. Anyone who has extensive experience with adsorption chromatography, for instance, would initially attempt to solve the problem in that way.

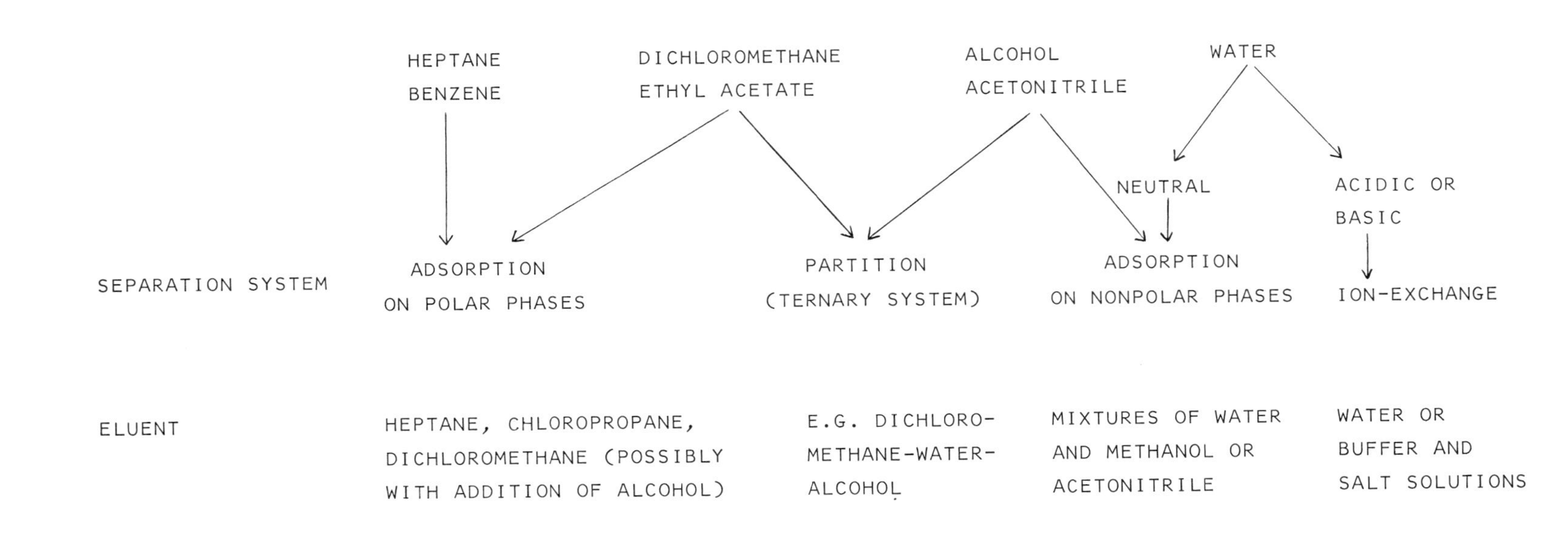

Fig.X.1. Selection of a separation system based on sample solubility

Table X.1 summarizes some selection criteria based on empirical observations, assuming that at least structural parameters are known [1]. For the separation of relatively nonpolar substances that differ in the type or position of the substituents or functional groups, adsorption systems appear to be the most promising. Positional isomers at double bonds (e.g., cis-trans isomers) are also excellently separated on such systems. The preferred eluents with polar stationary phases are the aliphatic hydrocarbons, chloropropane, methylene chloride, or chloroform, and mixtures of these. The polarity may be increased by adding a few percent of ethyl acetate, which has a relatively low UV absorbance. By reducing the specific surface area of the stationary phase, for example, by using silica gels with low specific surface areas or PLB, complete elution can also be achieved with the less polar solvents (cf. Fig.V.1).

If the sample components are still not eluted with the nonpolar solvent mixtures, a polar component (e.g., water) may be added up to saturation or ternary mixtures (e.g., methylene chloride/alcohol/water) may be used to convert the adsorption system continuously into a partition system. As detailed in Chapter VII, a liquid stationary phase is built up in the pores of the support as a result of the preferential adsorption of the polar eluent component. The rate of formation of the liquid phase depends on the support properties and the eluent composition. The constancy of the phase composition should be verified by frequent sample injection to determine the k' values or relative retentions. Partition chromatography occupies an intermediate position between adsorption and RP systems.

In principle, many separations achieved on polar stationary phases or by partition chromatography may also be performed on nonpolar phases. However, limited sample solubility in the required polar eluents often restricts the applicability of this approach. Nevertheless, RP chromatography yields superior separations of the members of an homologous series. The relative retentions of compounds differing by a CH_2 group are frequently $>$ 1.2.

Optical isomers have only been separated in the form of diastereomeric pairs (cf. Fig.VI.24), which generally poses no problems. Optically active stationary phases are a prerequisite for the resolution of racemates. Up to now, such phases are known only in the form of cellulose derivatives for classical column chromatography [2,3], and are unsuitable for HPLC. However, great efforts are being made to solve this problem chromatographically.

Table X.1. Selection of the separation system on the basis of sample structure

Structural parameter	Chromatogtaphic technique				
	Adsorption		Partition	Ion exchange[a]	Exclusion
	polar phases	nonpolar phases			
Molecular size	+	+	+		++
Isomers a) chain-ring	(+)	+	+		+
b) Branching for the same C-number	(+)	++	(+)		(+)
c) Steric (cis-trans)	++	++	+		
d) Optical	(?)	(+)	(?)	-	-
e) Number > = <	++	+	+		
f) Position of the > = <	+	+	+	(+)	
Homologous series	+	++	++		+
Substituents Number and position a) Nonpolar, e.g., alkyl, halogen	++	++	++		+
b) Moderator polar, e.g., nitro, carbonyl, ester	++	+	++		
c) Polar phenols, alcohols, amides, amines	+	++	++	(+)	
d) Strongly polar acidic or basic ionizable groups	(+)	++	++	++	

[a] only for purely aqueous systems

The applications of *ion-exchange chromatography* are relatively limited because they are generally confined to purely aqueous systems in which only ions can be separated. This can be extended, however, to substances that readily form reversible complexes with the ions bound to the exchanger (ligand exchange). In addition, there is the possibility that sorption occurs on the organic matrix of the exchanger, analogous to the reversed-phase mechanism. The addition of organic solvents to aqueous eluents leads to their demixing and to the formation of a partition system. If both of these separation potentialities were to be considered, the applications of ion-exchange chromatography would be more extensive and versatile than is evident from Table X.1.

Exclusion chromatography is primarily useful for separations based on molecular size, but exclusion effects may appear in all previously described systems. In polar eluents electrostatic forces (Donnan potential) may hinder even small molecules in penetrating the pores, even though this would not be expected from their molecular size alone.

When little is known about the sample composition, it may be advisable to determine its molecular weight distribution by EC. It should be noted that especially the polar polymers (e.g., proteins) tend to sorb irreversibly and go undetected, or tend to alter the separation characteristics of the column.

With all techniques the question occasionally arises whether all of the injected sample components have emerged or whether they are so strongly retarded that they are lost in the baseline noise. This is very difficult to answer. One approach is to repeat the separation using different systems. It is easiest to start by increasing the elution strength and observing changes in the position and number of the eluted peaks. A conclusive answer whether even the strongly retarded sample components have emerged can only be given after the separation is repeated in a system whose phase behavior is reversed. Thus, if one starts with a polar stationary phase such as silica gel, highly polar sample components may be retained. On the other hand, on an RP system with nonpolar stationary phase these components would be barely retained, thereby revealing their presence. The same holds, of course, for the opposite situation, where a change from an RP system to adsorption chromatography is recommended.

One should make it a rule to employ at least two completely independent separation systems for samples whose composition is largely unknown. Only then is a definitive answer concerning the composition and identification of the components possible.

To accelerate the *selection* of a suitable system, gradient elution could be used. This approach provides a rapid survey of the number and approximate polarity of the components present, the latter being based on their elution times and the gradient composition. This facilitates the selection of the eluent composition for an isocratic separation. Because of the constant possibility of ghost peaks caused by the enrichment of eluent impurities on the stationary phase, due care should be exercised in interpreting the chromatogram. An additional isocratic separation is always to be recommended. On polar stationary phases, the gradient should be initiated with of a nonpolar eluent (e.g., heptane) followed by a more polar one (e.g., methylene chloride), and finally with one containing a component that acts as a displacing agent (e.g., 10 - 50% isopropanol). Of course, this approach can be shortened. With nonpolar stationary phases the gradient is started with water and is gradually changed to methanol. Ghost peaks may also occur in this case.

The transfer of the results of classical column chromatography to HPLC presents no difficulties. However, this transfer refers to the separation system, and not to the exact retention data. Since the properties of the support play a significant role even in partition systems, any variations in its pretreatment may lead to increased deviations.

The transfer of thin-layer chromatographic (TLC) data to column methods is more difficult. The direct conversion of R_f values from TLC to the corresponding retention parameters (k' values, retention volumes) of HPLC is unreliable even through the R_f and k' values are formally related as follows:

$$R_f = \frac{1}{1 + k'} \, .$$

However, this relationship holds only under identical equilibrium conditions. In a column the phase ratio remains constant, but in TLC the volume of the mobile phase decreases toward the solvent front. Similarly, the phase composition in a column is constant, whereas in TLC a demixing (frontal analysis) of the developing solvent (eluent) is superimposed on the separation of the sample components. The mobile phase composition varies between the starting line and the solvent front [4,5], so that frequently several fronts appear. The migration rate of the solvent front (flow rate of the developer liquid) diminishes with increasing distance from the point of immersion into the solvent reservoir. Hence, it is impossible to operate isocratically in TLC, which is a precondition for the validity of the above equation. For the same reasons, the plate heights or numbers of TLC, which, of course, can be evaluated from the size of the spot, are not directly comparable to those of column chromatography. (H values are defined only for constant eluent velocity and phase ratio, isocratic conditions, etc.) In addition to the differences in the properties of the stationary phases, the binders used in TLC may exert further effects.

If it is assumed that the above equation is valid (even though is unacceptable for the reasons given), then R_f values of 0.9, 0.5, and 0.1 would correspond to k' values of 0.1, 1.0, and 9.0, respectively. These TLC separation conditions ($0.1 < R_f < 0.9$) can be achieved on a column ($0.1 < k' < 10$). The separation of 5 - 10 spots which is attainable on the upper half of a TLC plate ($0.5 < R_f < 0.9$) can be easily reproduced with about 1000 - 2000 plates on a column. On the lower half of the plate ($0.5 < R_f < 0.1$) about the same number of spots can be resolved, but because the corresponding k' region of a column is broader ($1 < k' < 10$), the column would evidently be superior to TLC.

For all of these reasons it is not surprising that difficulties arise in attempting to transfer the TLC results to the column mode. In addition, the R_f values obtained depend on the developing technique used (linear or circular development, in a saturated tank, sandwich chamber, or streaming system). A qualitative transfer from a TLC system with a nonpolar single-component solvent to a column system is relatively straightforward, although the water content of the layer, which is governed by the humidity, should be matched to that of the column (e.g., with an MCS). The separation can then be optimized by minor adjustment of the eluent polarity.

To transfer TLC conditions using multi-component solvents of very diverse polarities is much more difficult inasmuch as the solvent demixing alone prevents accurate estimation of the solvent and phase composition that are required to characterize the spot migration. It was shown [6] that using a two-component developer to separate various azo dyes yielded different results with the saturated tank and sandwich chamber techniques, and that these also differed from those obtained in column chromatography with the same eluent. It is obvious that the transferability of TLC separations to columns diminishes with an increasing number of TLC solvent components. Recently, TLC plates have become available [7] for the broad field of chromatography on nonpolar phases. However, the developer will rise only if it wets the layer and - neglecting forced flow - this will only occur if the solvent contains more than about 50% of the organic component. Hence, RP column chromatography with a predominantly aqueous eluent is difficult to reproduce on TLC plates.

To supplement the column chromatographic results, TLC equipment should be available in every HPLC laboratory. A TLC separation will

Table X.2. Separation Systems

	Principle of Separation				
	Adsorption (stationary phase)		Partition	Ion exchange	Exclusion
	polar	nonpolar			
Especially suitable for	Nonpolar to moderately polar neutral organic compounds that are soluble in nonpolar, water-immiscible solvents	Polar to moderately polar substances. Nonpolar substances only if they are water or alcohol soluble	All types of organic compounds	Dissociable substances such as amino acids, nucleic acids, alkaloids, inorganic salts	Separations based on molecular size
Stationary phase	Silica, sp. surface area > 50 m^2/g; alumina, active	Silica with chemically bonded alkyl groups (C_1 - C_{18}) (possibly activated charcoal)	Polar or nonpolar solids capable of absorbing at least one eluent component	Resin ion exchangers or those bonded to inorganic solids	Porous solids with a definite pore size distribution
Eluents (in order of increasing elution strength)	Alkanes, chloroalkanes, methylene chloride, chloroform, ether, acetonitrile, alcohols	Water, methanol, acetonitrile, higher alcohols, tetrahydrofuran, dioxane	Mixtures of at least two, preferably three, components (ternary mixtures) of different polarities, only partially miscible with each other	Aqueous, preferably buffer or salt solutions	Eluents that prevent interaction between sample and support surface
Elution rate may be accelerated by	Increase in eluent polarity; increasing water content, stepwise or continuously (gradient elution)	Decrease in eluent polarity, methanol-water or acetonitrile-water mixtures cover almost the entire range of potential separations	Increased sample solubility in the mobile phase	Increase in the ionic strength; change in pH; addition of ions of the same charge; complexation	Cannot be accelerated
Sample structure; retention increases with	Increasing polarity of the functional groups; dipole moment of sample, number of functional groups, molecular size	Decreasing sample solubility in water, molecular size, increasing number of nonpolar substituents, the alkyl chain length	Increasing sample solubility in the stationary phase	Acid or base strength; decreasing ionic radius	Decreasing molecular size. The separation by definition, is finished when the unretained substances (the smallest unretained molecule) is eluted. (Dead time or dead volume of the column)

readily establish whether all sample components can be eluted with the selected eluent (no sample remains at the starting point). This can be established relatively simply with the multitude of available detection reagents. Selective spray reagents can furnish additional information about the sample composition.

Table X.2 summarizes the characteristics of the separation systems discussed and gives the rules for selecting the system and for modifying the retention behavior of the sample.

References Chapter X

1. Hesse, G.: Chromatographisches Praktikum. Frankfurt/Main: Akad. Verlagsgesellschaft 1968
2. Lüttringhaus, A., Hess, U., Rosenbau, H.J.: Z. Naturforsch. *22b*, 1296 (1967)
3. Hesse, G., Hagel, R.: Chromatographia *6*, 277 (1973)
4. Geiss, F.: Parameter der Dünnschicht-Chromatographie. Braunschweig: Vieweg 1972
5. Engelhardt, H., Engel, B.: Chromatographia *1*, 490 (1968)
6. Soczewinski, E., Kuczmierczyk, J.: Chromatogr. *150*, 53 (1978)
7. Kaiser, R.E., Rieder, R.: J. Chromatogr. *142*, 411 (1977)

Chapter XI

Special Techniques

A. Preparative Chromatography

Preparative separations encompass the range from about 10 mg to a few grams of sample and can be performed with currently available instruments designed for analytical applications. Moreover, the range of 10 - 1000 mg is sufficient for further characterization by many modern research techniques.

A prerequisite for preparative separations is a good analytical separation of sample quantities that fall within the linear capacity region, i.e., a loading of 10^{-4} to 10^{-3} g sample per gram of adsorbent. The load capacity is defined as the maximum amount of sample that can be applied without impairing the separation efficiency. On the usual analytical columns (3 - 4 mm i.d., 30 cm long), quantities of about 1 mg can be separated without further ado, which is already a "preparative" amount for many expensive natural products. Analytical separations, with sample quantities within the load capacity, may be optimized by adjusting the separation conditions so that the resolution, i.e., the distance between the peaks, is very high. (This is achieved, of course, at the expense of the analysis time). Provided the peaks are sufficiently separated from each other, the sample quantity can be increased, and the emerging broader zones will not overlap because the resolution is so high. In this way, "preparative" amounts of 5 - 100 mg may be separated on ordinary analytical columns. It should be noted that the retention times decrease as the sample size is increased. With more sample, the elution system may eventually be converted to a displacement system (cf. Chapter I), with each succeeding peak displacing the preceding. There will then be no pure eluent zone between the individual peaks.

The throughput may be increased by enlarging the column diameter, but doubling the column cross-section requires quadrupling the solvent delivery rate of the pump to maintain the same linear carrier velocity. The pumps used in HPLC generally have a maximum output of 10 - 15 ml/min, which restricts the enlargement of the diameter, although one of 8 - 10 mm poses no appreciable difficulties and yields the same efficiency as analytical columns. Even wider columns (up to 25 mm) have been employed successfully [1-5] with slightly modified analytical instruments. The abrupt transitions from the usual capillary tubing (0.5 - 1 mm i.d.) to the full column cross-section are accompanied by flow disruptions at the column inlet and outlet. This can be mitigated by means of cone-shaped (120 - 150° angle) transitions packed with the stationary phase [6].

The sample may be introduced either via syringe or valve, but the latter is preferred when dealing with larger volumes (10 - 50 ml) of dilute solutions, as additional band broadening is negligible in preparative work. However, the introduction and uniform distribution of the sample over the entire column cross-section does present occasional difficulties. Most significantly, the rule of thumb that the sample solution should be as concentrated as possible does not hold. It was shown [7] that the introduction of a more dilute solution of the same quantity of sample resulted in less band broadening than with the concentrated one. This effect was attributed to local overloading of the column.

For the above reason and due to the occasionally limited solubility of the sample components in the eluent selected, a relatively large volume must be introduced. The sample volume does not begin to affect the peak shapes until it exceeds the column standard deviation (in volume units) which results from the mixing processes within the column [8]. If a 20% increase in the H value can be tolerated, the sample volume may be increased to about 1.5% of the empty column volume. (For the usual 30 cm, 4 mm i.d. analytical columns this corresponds to sample volumes of 50 - 150 μl). If sample resolution can be achieved even with a 50% increase in H, 3.0 - 3.5% of the column volume may be injected [9].

The ordinary commercial detectors are generally too sensitive for preparative purposes. The sensitivity of photometric detectors can be lowered by reducing the path length. However, at a higher eluent throughput, fluctuations and increased noise are observed with some detectors. The outlet tube from a detector cell should be as short and wide as possible to avoid excessive pressure on the cell windows at high flow rates, which may lead to cracks. The eluent stream may be split by incorporating a drilled out T-connection after the outlet capillary, and not after the end of the column, to avoid "infinite diameter" effects.

For preparative separations there are two principal alternatives. One is scale-up of analytical separations and the use of ~ 10 μm particles, the other is utilization of moderately efficient columns packed with 50 μm or larger supports. Both approaches possess advantages and disadvantages.

The larger particles are relatively easy to pack, are quite inexpensive, and place no particular demands on the equipment (pump). Their efficiency, however, is low (optimally 1000 plates/m) but can be improved by recycling.

The smaller particles (10 - 20 μm) yield high-performance columns (10 000 - 30 000 plates/m) capable of resolving complex mixtures. Because of the small peak volumes the concentrations of the separated components remain high, thereby facilitating removal of the eluent. Their main drawbacks lie in their high cost and the relatively expensive pumps required for packing and for eluent delivery.

The throughput, which is particularly important for preparative separations, is independent of the particle size of the stationary phase, and is only a function of the column cross-section and eluent flow rate.

The selection of the appropriate preparative apparatus depends upon the particle size to be employed. For highly efficient columns, repetitive manual or automatic separations will be the first choice.

Instruments for automated preparative separation with fraction collection and repetitive sampling controlled by a programmer are commercially available. These instruments not only permit collection of the individual fractions, but also automatically introduce a new sample after finishing a predetermined analysis cycle, and combine the new fractions with the corresponding previous ones. This, however, requires constant separation conditions for a long period of time; the associated technical problems will not be dealt with further here. This repetitive approach using analytical or somewhat larger (~ 10 mm) columns appears to be the most convenient solution at the moment, since most analytical instruments can be readily modified for this purpose.

An alternative, that reduces the amounts of eluent and support, is the application of the recycling method [10]. This approach is frequently used in exclusion chromatography to increase the plate number. The column may be significantly more heavily loaded in this case because the sample is diluted continuously. Theoretically, recycling can be continued until the fastest peak catches up with the slowest. It is important to minimize mixing (and hence band spreading) in the pump, i.e., between the end and head of the column.

A commercial preparative liquid chromatograph using columns packed with large particle sizes has been designed to maximize the throughput at minimum cost. This is achieved at the expense of resolution. The columns (5 cm i.d.), made of polyethylene and containing inexpensive, coarse (50 - 100 μm) silica gel, can be easily inserted as a cartridge into the column holder. The optimal packing density is improved by radial compression between the column holder and the column cartridge, which is also supposed to improve the efficiency. The columns are not very expensive and are easily interchangeable. Due to the low pressure drop required, high outputs (50 - 500 ml/min) can be attained with a relatively simple pump. This concept permits a few grams of sample to be separated easily. Prepurification steps and the removal of non-eluting components (e.g., proteins in biological materials, etc.) are not necessary. This approach appears to be especially adapted to the separation of relatively few components (e.g., after preparative-organic synthetic steps, etc.).

B. Qualitative Analysis

For all chromatographic processes the unambiguous identification of substances requires agreement between its net retention times or, preferably, k' values and those of a known substance, measured with at least two different separation systems. The k' values are superior for tabulation and calculations because they are independent of the eluent velocity and column cross-section. Since minor variations in the eluent composition may exert pronounced effects on the analytical results, it is advisable to define the chromatographic conditions via k' values or relative retentions of standard substances. The relationships discussed under various separation systems (e.g., the dependence of the k' values on the C-number, etc.) are also useful adjuncts for qualitative analysis. The systematization of sample retentions, at least for chromatography on nonpolar phases, along the lines of the Kovacs Indices in gas chromatography would also be very beneficial for the identification of unknowns. A relatively simple alternative consists of collecting the emerging peaks after the detector and conducting further identification tests off-line.

C. Quantitative Analysis

Quantitation in HPLC is similar to that of other chromatographic techniques. Basically, the area under the peak is proportional to the amount of substance injected. Which method is chosen for the determination of the area depends on individual preference. As in GC [11,12], these include the peak height, peak height times width at half-height, the area, integration via disc or electronic integrators or a computer [13]. The errors lie between 1 and 5%, and in the case of graphical evaluations depend on the skill of the individual.

The sources of error are known, in part, from gas chromatography. Sample injection via syringe has certain shortcomings. As a result of the high back pressure, it may be difficult to inject identical amounts of sample. Especially after prolonged use, there is the risk that the majority of the sample will escape along the plunger. For precise quantitative work, particularly for serial analyses, valve injection is preferred because of its better reproducibility at high inlet pressures. To eliminate errors from syringe injection, the internal technique should be used. This technique consists of adding an exact amount of a substance known to be absent from the sample, preferably one that will elute at a "vacant" position on a chromatogram. Injection errors can then be corrected on the basis of the peak area of this standard. Errors attributable to the chromatographic separation process frequently result from incomplete sample elution. Hence, serious error may result from normalizing on the basis of the eluted peaks. This source of error can be eliminated by performing the analysis with two completely independent chromatographic systems.

The response of HPLC detectors is concentration specific [14,15]. A recorder shows the variation in the concentration with time. In quantitation by evaluation of the area the concentration is integrated as a function of time ($\frac{g}{cm^3} \cdot sec$). To obtain the mass (g), the area must be multiplied by the flow rate ($\frac{cm^3}{sec}$). Therefore, quantitation by area evaluation cannot be more accurate than the constancy of the flow rate. This includes the long-term constancy during the entire analysis as well as the short-term constancy during the elution of a peak. The flow rate of all commercial instruments is guaranteed to be constant to 1%. In quantitation based on peak height, it is essential to establish that the k' values or retention volumes remain constant.

The response of LC detectors is substance-specific. Consequently, a separate calibration curve must be constructed for each substance. Moreover, when using UV-detectors the molar absorptivities quoted in the literature should not be used, even when working at the same wavelength. The spectral band width of the ordinary UV detectors for HPLC lies between 5 and 15 nm, whereas the molar absorptivities are usually measured with 0.5 - 1 nm band widths. Therefore, the pertinent absorptivities for the quantitation of chromatograms vary with the detector used or its spectral band width; they also depend on the band shape and are usually lower. For example, for benzene ε_{max} = 215 at 255 nm [16], whereas measured at 254 nm using a 10 nm spectral band width, the molar absorptivity is found to be about 100.

The relatively broad spectral band width restricts the linearity of LC detectors. The steeper the absorption band, the greater are the deviations from the Beer-Lambert Law [17]. Such deviations are negligible only for substances with very flat absorption maxima. Because of these deviations the linear range of UV detectors is restricted to about $2 \cdot 10^2$ concentration units. If the range is to be extended to 10^3 concentration units, deviations from linearity of 10% or more must be tolerated.

Because of the variation of absorbance with temperature (especially when measuring on the shoulder of an absorption band), both column and detector cell should be maintained at constant temperature for quantitative analyses. Furthermore, all the common precautionary measures in photometry [17] should be observed.

D. Trace Analysis

Owing to the limited sensitivity of LC detectors, one is frequently faced with the task of determining traces of a particular substance in a mixture. Even if the original concentration is adequate for a determination, it will be much lower after separation, due to the chromatographic dilution process. In order to enable detection more sample should be injected. This poses no difficulties especially for very dilute solutions, since the column is not overloaded. As a rule, the sample volume may be up to about 2% of the column volume without increasing the band broadening noticeably. The sample volume may be readily increased further if the substance being determined has a k' value greater than about 2 and if the prior-eluting components are not to be measured. Such a sample (with k' > 2) is concentrated in the stationary phase by being retarded relative to the eluent and migrates as a narrow zone through the column.

Thus, the column can be used to concentrate trace quantities or dilute samples. By selecting a system in which the components of interest are strongly retarded, they can be concentrated at the top of the column. This is carried out by repeated injections of the diluted sample solution or, in the extreme case, by pumping the sample solution itself through the column. By increasing the elution strength of the solvent, the enriched components may be caused to migrate down the column. This sample enrichment is similar to adsorptive filtration (cf. Chapter I) and becomes particularly useful for the determination of minute quantities in large volumes. For aqueous solutions a reversed-phase system should be employed, and for nonpolar solvents an ordinary adsorption system with a strongly polar stationary phase is appropriate.

A short, steep gradient (small gradient volume) may also be used to advantage for sample concentration. If this is not possible or the desired sensitivity is unattainable, the detection sensitivity may still be increased by minimizing dilution during the separation process. Other means include improvement of the detector properties through noise reduction, optimization of the wavelength and of the cell path length, etc., and will not be discussed further here.

A certain number of theoretical plates is necessary for every separation. Inasmuch as dilution is a function of the column length or residence time in the column, (k' value of the sample) only the shortest column possessing the necessary plates should be used for trace analysis. Since the plate height also decreases with decreasing particle size, the smallest possible particles should be used to achieve the required number of plates. Because the plate height is a function of the linear velocity, the analysis should be performed at the flow rate that corresponds to the minimum plate height (H_{min}, U_{min}, N_{max}). In this way the separation efficiency is optimized at the expense of the analysis time. The chromatographic conditions should be adjusted to achieve low k' values, as high ones result in excessively broad, flat peaks. Since sample dilution decreases with decreasing column diameter, the narrowest column, that can still be packed well, should be employed. A detailed discussion of the effects of these parameters on the detection limit may be found in the literature [18-20].

In summary, the shortest column packed with the smallest particles should be used for the detection of trace components. The internal diameter should be as narrow as possible and the carrier velocity should be in the vicinity of u_{min}. The system should be adjusted to obtain the optimum k' values (1.5 - 4). When there is no concern about overloading the column (extremely dilute solution) the sample volume can be increased until broadening of the sample peak becomes apparent.

A different approach should be used if trace components are to be detected in a sample, that is available in unlimited quantities. In such cases the column diameter is unimportant [21]. The amount of sample can be increased up to the limit of the load capacity, perhaps even in excess, particularly if the selectivity of the system is ample for the resolution of the trace components. The difficulties associated with this type of trace analysis may be minimized by choosing a separation system in which the traces of interest elute before the major constituents.

References Chapter XI

1. Wolf III, J.P.: Anal. Chem. *45*, 1248 (1973)
2. Wehrli, A.: Z. Anal. Chem. *277*, 289 (1975)
3. Larmann, J.P., Williams, R.C., Baker, D.R.: Chromatographia *8*, 92 (1975)
4. Attebery, J.A.: Chromatographia *8*, 121 (1975)
5. Godbille, E., Devaux, P.: J. Chromatogr. *122*, 317 (1976)
6. Beck, W., Halâsz, I.: Z. Anal. Chem. *291*, 340 (1978)
7. De Stefano, J.J., Beachell, H.C.: J. Chromatogr. Sci. *10*, 654 (1972)
8. Wehrli, A., Hermann, U., Huber, J.F.K.: J. Chromatogr. *125*, 59 (1976)
9. Beck, W., Halâsz, I.: Z. Anal. Chem. *291*, 312 (1978)
10. Conroe, K.: Chromatographia *8*, 119 (1975)
11. Kaiser, R.: Gas-Chromatographie. Vol. IV: Quantitative Bestimmung. Mannheim: Bibliograph. Inst.
12. Ettre, L.S., Zlatkis, A. (Eds.): Practice of Gas Chromatography. New York: Interscience 1967
13. Karger, B.L., Barth, H., Dallmeier, E., Courtois, G., Keller, H.E.: J. Chromatogr. *83*, 289 (1973)
14. Halász, I.: Anal. Chem. *36*, 1428 (1964)
15. Halász, I., Vogtel, P.: J. Chromatogr. *142*, 241 (1977)
16. Silverstein, R.M., Bassler, G.C.: Spectrometric Identification of Organic Compounds. New York: Wiley 1964

17. Kortüm, G.: Kolorimetrie-Photometrie and Spektrometrie. 4. Aufl. Berlin-Göttingen-Heidelberg: Springer 1962

18. Meijers, C.A.M., Hulsman, J.A.R.J., Huber, J.F.K.: Z. Anal. Chem. *261*, 347 (1972)

19. Karger, B.L., Martin, M., Guiochon, G.: Anal. Chem. *46*, 1640 (1974)

20. Kirkland, J.J.: Analyst *99*, 859 (1974)

21. Halász, I., Endele, R., Asshauser, J.: J. Chromatogr. *112*, 37 (1975)

Chapter XII

Purification of Solvents

One of the most frequent causes of poor reproducibility in LC separations is impurities in the solvents used as eluent. Even minute traces of impurities may lead to spurious results due to their enrichment on the column, as detailed in Chapter VI. This occurs primarily during adsorption-chromatographic separations where small traces of polar contaminants, such as water, alcohol, etc. alter the properties of the system.

In principle, all eluents should be distilled before use. This is imperative if the samples are to be recovered after separation, since nonvolatile eluent impurities could contaminate the sample. In gradient elution the impurities in nonpolar eluents may be concentrated on the column and, in going to stronger eluents, be eluted as sharp, but false peaks. This is illustrated in the lower chromatogram of Fig.XII.1. The upper chromatogram was obtained in a gradient elution run with heptane, containing substantially fewer impurities and demonstrates the ameliorative effect of distillation plus adsorptive filtration over alumina.

This method of purification [1-5] not only removes the polar contaminants, including water, but at the same time furnishes an eluent with a markedly improved UV transparency. Methylene chloride and chloroform, whose transparency limits lie near the most often used wavelength (254 nm) of UV detectors, must frequently be purified in this way before use. This also markedly improves the linear response range in them.

The *purification* of solvents is carried out via classical column chromatography: a tube (e.g., 2 - 5 cm i.d., 40 - 150 cm long) is filled with a highly active adsorbent such as alumina or silica and immediately covered with the eluent. The initial portions of eluent issuing from the column are collected separately. They are not pure enough, but may be reprocessed. The middle and main fractions represent pure solvent. The yield depends on the adsorbent activity (specific surface area), the solvent polarity, the amounts of impuri-

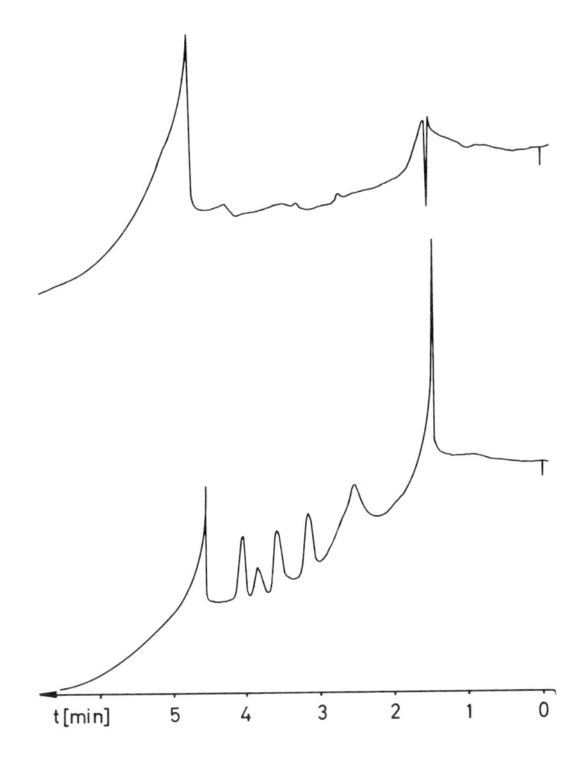

Fig.XII.1. Purification of solvents. Gradient elution without sample. n-heptane → methylene chloride. Lower: column flushed 15 min with distilled n-heptane, gradient to methylene chloride. Upper: column flushed 15 min with n-heptane that had been additionally passed over alumina. Experimental conditions: silica Si 100; column: 30 cm, 4 mm i.d.; F = 4.6 ml/min; p = 70 atm

ties and their polarities. The following are guideline values that may be kept within or exceeded. The process should be standardized by a single determination of the breakthrough volume of the impurities for a given adsorbent and eluent. With 100 g of alumina, about 150 - 600 ml of aliphatic hydrocarbons (pentane, hexane, heptane, etc.) can be purified. Silica gel has a higher capacity due to its greater specific surface area. The best results were obtained with a mixed column (alumina and silica) [4]. To regulate the flow rate the column should be equipped with a stopcock, which, however, must not be greased.

Of course, the throughput for polar eluents such as methylene chloride, chloroform, etc. per unit weight of adsorbent is smaller (200 - 400 ml/100 g alumina). With pre-dried eluents the purification capacity increases accordingly.

This method yields dry eluents which are also free of polar impurities (alcohols, aldehydes, ketones, etc.).

Eluents that fall below ethyl acetate in the eluotropic series (Table VI.2) cannot be dried in this way. They may be dried, but not purified, over molecular sieve 3 Å. Eluents are conveniently stored over the molecular sieves, but their large particle size (> 1 mm) prevents normal column purification. Solvents dried in this way contain fines which may damage pump valves. Very dry eluents are hygroscopic. Should it be necessary to work with them, it is absolutely essential to exclude moisture and to store them over freshly regenerated and dried molecular sieves.

Solvents for chromatography on nonpolar phases are far more difficult to purify. Deionized water is sometimes not usable, and even water distilled in a quartz still often contains organic compounds, especially if they are steam-distillable. In crucial cases, particularly for gradient elution, purification by frontal analysis, as described above, but using nonpolar stationary phases, is essential. Methanol and acetonitrile are commercially available in analytical and spectroquality grades, but both grades may differ in their "chromatographic" purity. If measurements are not to be performed at low wavelengths (< 240 nm), the analytical grade is frequently to be preferred, to a large extent because it is substantially cheaper than the spectroquality material. Solvents especially purified for chromatography (including water) are commercially available from various suppliers.

References Chapter XII

1. Wohlleben, G.: Angew. Chem. *67*, 741 (1955)
2. Wohlleben, G.: Angew. Chem. *68*, 752 (1956)
3. Hesse, G., Schildknecht, H.: Angew. Chem. *67*, 737 (1955)
4. Hesse, G., Engelbrecht, B.P., Engelhardt, H., Nitsch, S.: Z. Anal. Chem. *241*, 91 (1968)
5. Engelhardt, H., in: Zief, M., Speigths, R.M. (Eds.): Ultrapurity. New York: Dekker 1972

Subject Index

Anleitungen für die chemische Laboratoriumspraxis

Herausgeber:
F. L. Boschke, V. A. Fassel,
W. Fresenius,
J. F. K. Huber, E. Pungor,
W. Simon, T. S. West

Band 13: K. Cammann

Das Arbeiten mit ionenselektiven Elektroden

Eine Einführung

2., überarbeitete und erweiterte Auflage. 1977.
65 Abbildungen, 15 Tabellen. XII, 227 Seiten
ISBN 3-540-07947-5

Inhaltsübersicht: Grundlagen der Potentiometrie. – Elektrodenpotentialmessung. – Ionenselektive Elektroden. – Meßtechnik bei ionenselektiven Elektroden. – Analysentechniken unter Benutzung ionenselektiver Elektroden.

"The description of the basic principles regarding ion-selective electrodes and their possible application to the solution of practical problems is the main purpose of *Cammann's* book. Its first part deals with the theoretical side of the subject and with the explanation of conventional terms; the second is devoted to the problems that may be faced by the scientist or technician during the performance of precise and reliable measurements by such electrodes...

The appendix that completes the volume contains tables indicating activity and temperature coefficients and also other useful data. The book is not only for scientists active in a variety of fields, but will also be an excellent guide for students who intend to learn how to perform practical analytical measurements."

Bioelectrochemistry and Bioenergetics

The English translation of this book, **Ion Selective Electrodes,** will appear in 1979.

Springer-Verlag
Berlin
Heidelberg
New York